Mineral Resources

PROBLEMS IN MODERN GEOGRAPHY

KENNETH WARREN
in

Mineral Resources

A HALSTED PRESS BOOK

JOHN WILEY & SONS
New York

© 1973 by Kenneth Warren

Published in the U.S.A. by
Halsted Press, a Division of
John Wiley & Sons, Inc. New York

ISBN 0 470 92116 1

Library of Congress Catalog Card Number: 72 11476

Printed in Great Britain

To my Parents

While corn is springing from the earth above,
what lies beneath is raked over like a fire,
and out of its rocks comes lapis lazuli,
dusted with flecks of gold.
No bird of prey knows the way there,
and the falcon's keen eye cannot descry it;
proud beasts do not set foot on it,
and no serpent comes that way.
Man sets his hand to the granite rock
and lays bare the roots of the mountains;
he cuts galleries in the rocks,
and gems of every kind meet his eye;
he dams up the sources of the streams
and brings the hidden riches of the earth to light.
But where can wisdom be found?

Job xxviii, 5–12, *New English Bible*

Contents

Maps and Diagrams

Mineral Resources

List of Illustrations

Mineral Resources

Preface

All that one sees on the earth's surface is made of or from either plant life or minerals. Ultimately all of it is derived from the minerals of the earth. The buildings of our towns, railway lines, the tarmac of roads, our implements and machinery, ships, planes and most of the energy which moves them, a large proportion of clothing and other fabrics, are all made from material drawn from open pits, shafts or bore holes in the surface of the earth. The material apparatus of a modern industrial society becomes ever more complex, and as standards of living rise so does mineral consumption. If the three quarters of the human population living in the undeveloped world begins to share in the affluence which we enjoy, growth in mineral consumption will rocket.

Throughout the thirty years following the First World War the United States was the world's only major fully-fledged affluent society, already in the development stage of high mass consumption. In that period alone it consumed more minerals than the rest of the world had done through the whole of history. By 1960 average annual American per capita consumption of non-power minerals was valued at $40 as compared with an average of less than $1 in the underdeveloped world.[1] When one adds the compound interest of world population growth, and the continuing advance of the already industrialized countries to the backlog of the developing world, the prospect for expansion in mineral demand becomes daunting.

One 1970 estimate suggested a general world increase in mineral consumption of 5 per cent a year. Another forecast looked to a trebling of world demand by the end of the century, and a 1968 conference in Sacramento, California was told that mineral consumption may double by 1985 and double again by A.D. 2000. Assuming a steady improvement in the position of the Third World countries yet other commentators have suggested a growth rate which, at least in the early years, would involve a doubling of present world demand every seven or eight years.[2]

Growth on anywhere near these varied scales has wide ranging development implications. The demand for capital will be immense – for transport routes, often into highly inhospitable areas, and the construction there not only of mines, concentrators and perhaps smelters, but of townships, schools, hospitals and so on. There will be migration of labourers and still more of skilled men. Traditional agriculture and indeed whole existing economies will be transformed with new economic and social relationships emerging. The need for geologists, mining engineers, metallurgists and a host of other specialists will increase. The advanced industrial countries will have to provide new refinery sites and fabricating plants, more office accommodation and above all fresh sources of finance. In turn, these nations must develop new relationships with independent, sometimes aggressively independent, governments in the major mineral producing countries of the developing world. Not only is economic growth multi-faceted but it is influenced by and has implications for conditions far beyond the narrowly economic. In short, modern, large-scale development of the earth's mineral resources is an important force transforming the earth's surface, the way of life of its populations and the relationships which bind them together.

Given a prospective sharp increase in mineral demand the future nature and distribution of that industry will be determined by the facts of world geology, the state of man's knowledge of them, the economics of mineral working, transport and processing and the company and political framework within which these economic factors work. Where major mineral expansion occurs old national and regional economies will be transformed and new relationships established between man and his environment. In societies increasingly conscious of the possible harmful effects of unco-ordinated change, these developments require planning. The following chapters explore these aspects and ramifications of the exploitation of world mineral resources.

Kenneth Warren, *Witney, Oxfordshire. January 1972*

1. Introduction

Until the early nineteenth century almost all human societies seem to have been built on variations of a single economic model. The mass of the people lived at subsistence level while a very small part of the population enjoyed most of society's material goods. If food supplies increased, the standards of the toiling proletariat would rise for a time but soon population growth would again reduce it to subsistence level. Such a society already depended on mineral production, not only for fine building stone, precious metals and gemstones for the adornment of palace, house and person, but for the base metals used by the masses in weapons, tools, ploughs and so on. By present standards per capita demand was very small, and transport was primitive, so that local sources of supply were usually adequate to meet all needs, but already there were some more long distance movements. In the empires of the ancient world caravans and coastal vessels carried stone or metal ores to meet the whim, the military and the economic needs of the powers of the state. Egypt quarried turquoise and copper ores in Sinai, and the gold of Punt and Ophir supplied Egyptian needs and those of Old Testament Israel. The ability to produce weapons of bronze widened the power of empires, and their strength increased still more with tools and weapons of iron. As it was difficult to attain temperatures high enough to smelt it, iron was still a rare and highly valued metal in the second millenium B.C., in the Middle East. Anittas, a Hittite king of at least as long ago as the eighteenth century B.C., was pleased to receive a sceptre and a throne of iron as tribute from a conquered city. Later the radius of mineral trade widened, the Phoenicians' traffic in tin from Cornwall, the 'Cassiterides', being an outstanding example. Even in the Ancient World mineral wealth was a source of revenue, economic power and an object of cupidity. Laurium in classical Greece, and that other ancient world of pre-Columban America, may be taken to illustrate the state and some of the effects of pre-industrial mining.

Laurium is twenty-five miles south of Athens near the Sunium peninsula. Its silver mines supplied the metal for the Athenian coinage and were one of the chief sources of the wealth of the state. Individuals or groups of citizens leased the mines, working them with the least intelligent or most vicious slaves, chained and almost naked. After the battle of Marathon in 490 B.C. Themistocles persuaded the Athenians to turn the mine revenues into shipbuilding, so laying the foundation for the great naval victory over the Persians at Salamis ten years later. Shortly after this the output of the mines declined, but in the first century B.C. Strabo reported that the mine tailings were being worked over for lead, a metal which had had little interest for Athens. And, much more recently, in the late nineteenth century, French and Greek concerns were again working the mines, this time for lead, manganese and cadmium.

The indigenous empires of the New World had not mastered the use of iron but their wealth in precious metals brought the covetousness of the Conquistadores to fever pitch. Hernando Cortes is said to have remarked: 'I and my companions have a complaint, a disease of the heart, for which gold is a certain cure'; and in Peru the cupidity of Pizarro and his men was not one whit less. On the *noche triste* (night of melancholy) when Cortes withdrew from Mexico City, many of his men died under Aztec attack by slipping off the causeway, weighted down with gold. In Peru the Inca, Atahualpa, offered a ransom for himself – to fill his prison with gold as high as he could reach. In return for the wealth of plunder, and that later drawn from the mines of the Andes and Middle America, Spain ruined the economy of its New World empire. Prescott, the great historian of the conquests, allowed himself a florid passage of comparison between the patterns of development in Anglo- and Latin America.

What a contrast did those children of Southern Europe present to the Anglo-Saxon races who scattered themselves along the great northern division of the western hemisphere! For the principle of action with these

2

latter was not avarice, nor the more specious pretext of proselytism; but independence – independence, religious and political. To secure this they were prepared to earn a bare subsistence by a life of frugality and toil . . . No golden visions threw a deceitful halo around their path, and beckoned them onwards through seas of blood to the subversion of an unoffending dynasty.[1]

Not only did the wealth of the Spanish New World help to frustrate more balanced, securely based economic development there, but it upset that of Spain itself. The influx of precious metals hastened the decay of Spanish pre-eminence, and as a result of trade, or leakages from the system due to the depredations of English or other pirates and privateers, affected the whole economy of Renaissance and early modern Europe. It therefore contributed to the new European balance of power and the later differences in levels of economic development between northern and southern Europe. However he would be a rashly bold man who maintained that this was the chief cause.

Much later new mines for base metals and coal helped to draw American settlement westwards, and the gold of the western cordilleras and then the precious and base metals of the Canadian Shield populated new wastelands. In the nineteenth century Australian gold camps attracted settlement from the sheep, cattle and grain country and the coastal agglomerations of people, and the gold of the Rand with copper further north brought settlers and the railway to the interior of central Africa.

All this suggests that there is much in the contention that mining has been the cutting edge for economic development, the touchstone for advancement of backward areas. This was the argument of a fascinating two-volume survey by the expert mining and metallurgical journalist Rickard, a correction of what he regarded as the under-estimation of the role of minerals in the development of civilization in H. G. Wells's *Outline of History*. Rickard's last words summed it up, 'Trade follows the flag, but the flag follows the pick.'[2] Earlier Keynes had argued that the rise of the German Empire owed more to coal and iron than to the Bismarckian policy of blood and iron, and in the late twenties and early thirties Stalin was justifying the harsh discipline of the Five Year Plan as essential for the laying of the mineral and heavy industry base which alone could defend 'socialism in one country'.

Two recently published statements, though originally made twenty or twenty-five years ago, sum up the extremes of points of view concerning 'mineral determinism'. On the death of the geologist H. H. Read, a 1952 paper was recently reprinted. In it he argued that the world pattern of wealth essentially reflects that of minerals. 'As I have often protested, the North Atlantic countries are not really populated by people especially virtuous or specially noble, industrious or gifted, or even especially acquisitive, they happen to be countries which had a geological history favourable to the formation of mineral deposits, and upon this lucky foundation has been built their industrial structure.'[3] At the beginning of January 1971 The Times published a letter which Field Marshal Lord Wavell wrote to a friend in 1947. In it he reflected on the situation and prospects of postwar Britain. 'Unless we can get back to something like our old standards of honesty, family morality, hard work and pride in craftsmanship, I do not feel that we shall maintain our position or regain our former prosperity, which was founded on the above qualities more than on other things: more, for instance, than on the fortuitous location of coal and iron ore in the British Isles.'[4]

There can be no gainsaying the very great significance of basic mineral resources in the advancement of standards of living and in military power. However, the long continuing controversy over the origins, and especially over the reasons for the localization of the industrial revolution in western Europe shows that the situation is extraordinarily complicated. The society, the men, and the mental framework or ethos of a people must be right as well as the mineral fields, and the picture becomes almost hopelessly confused when it is realized that none of the human factors are independent variables. Perhaps it may be suggested that in the early stages of development of today's industrially advanced nations mineral endowment was a vitally important advantage. At an advanced stage, modern commerce and its material base in low cost bulk transport have opened the whole world as a source of supply, so that it is the other human qualities, those about which Wavell wrote, which determine the growth rate of the G.N.P. Given the

4

established positions of the 'have' and the 'have not' nations this cannot be of much consolation to the latter. Too close a correlation of mineral wealth and economic and political power is clearly unacceptable.[5]

Everywhere mining has left a distinctive imprint on landscapes and economies. In classical times the preparation of the charcoal used in smelting and the sulphurous fumes from the operation itself helped deforest the area surrounding Laurium. Around the old mines of Spanish America vegetation is still poor as a result of the cutting of timber for fuel and overgrazing by mine animals. Spanish colonial America indeed provides an example in which the whole economy and society was sacrificed for mineral wealth. Under the 'mitta' system Indians were forced to labour in the mines with a resulting mortality estimated as ranging up to 80 per cent in the first year of employment. Irrigation works and the Inca roads decayed; the birth rate fell as Indian males were separated from their families. Population declined disastrously, in the case of Peru from an estimated 8 million in 1575 to less than 1.1 million at the time of the first fairly trustworthy enumeration of 1793.

Even though this is far and away the extreme case of the disastrous results of a preoccupation with minerals it is still undesirable to depend too heavily on revenue from them. A country with aspirations for securely based economic advancement must build up less transient sources of wealth. As one paper to a U.N. conference on minerals put it, funds from mineral sales '. . . should be converted to other forms of capital which are of equal or greater value, particularly into basic industrial installations, and into power, transportation and communication systems. The conversion of high grade resource assets into current living expenses by means of export can lead to tragedy.'[6] This points to a final and most important need in world mineral production and trade. A system must be devised to provide the mineral producing, poorer nations with a higher, more stable price for their products. This will be a major contribution to the removal of the iniquities of the wide range in world living standards.

2. Geological Considerations and Mineral Survey

The world geography of mining may be picked out on a map of patterns of production and consumption, of the distribution of mines and smelters, and the mineral flows which link them. It is the product of the interaction of a group of factors. The motive force for these evolving patterns is effective demand, that is need backed by purchasing power. Traditionally the demand has come from a multitude of consumers each responding in their turn to market opportunities, each swept up in the process of economic growth. Now, fewer, bigger manufacturing units are the immediate purchasers of the production of the mineral industries and government intervention is common. In a large section of the modern world, the communist bloc, with between one quarter and one third of world population and about 23 per cent of the world's land area, state intervention has gone further and mineral producers and consumers alike are given targets of planned demand. But whether it is a free enterprise or a command economy, there is competition for orders or for development funds between various sources of supply – home or overseas mines, established but expanding operations or wholly new prospects. Although there is a good deal of flexibility, and many alternative choices are possible, decision takers of whatever economic organization or political hue must take note of the facts of geology so far as they are known. These are quite literally the basic considerations in mining, tying the economics of the business to the earth's surface and giving rise there to patterns of production and flows of raw materials. The geography of mining is the result of the response of enterprise to the known geological facts, though conditioned by the availability of capital and labour, by political considerations, planning controls and, very importantly, by the legacy from past patterns.

6

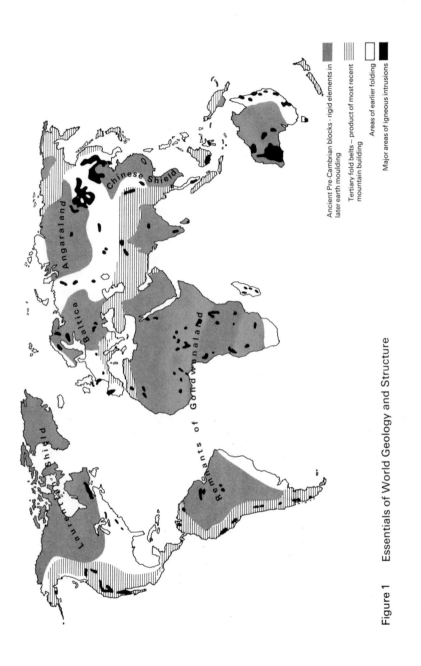

Figure 1   Essentials of World Geology and Structure

Ancient Pre-Cambrian blocks - rigid elements in later earth moulding

Tertiary fold belts — product of most recent mountain building

Areas of earlier folding

Major areas of igneous intrusions

Laurentian Shield

Angaraland

Chinese Shield

Baltica

Remnants of Gondwanaland

Mineral Resources

Some areas of the earth are mineral rich, others are not: in some, resources are pretty thoroughly evaluated while even today our geological knowledge of very large areas is sketchy, and that of the almost three quarters of the crust below the seas and oceans is slight except for a few quite inadequate indices – depths, the broad pattern of contours and a sampling of the superficial deposits of the bed.

Mineral Geology

Occasionally, and especially in the pioneer heroic days of mining which preceded the twentieth century, almost pure or 'native' metals were mined, as with the all-too-rare gold nugget or mass of copper. Normally the metallic minerals form chemical combinations with other elements and are now, necessarily, worked from deposits containing only a very low percentage of metal. Common combinations are oxides, carbonates, sulphides or chlorides. Each ore body will be surrounded by a non-mineralized zone, or one in which mineralization is so slight that in the current state of technology the material is not regarded as an ore. In mass mining from open pits and to some extent also in the selective mining of a lode or vein some of this waste or 'gangue'* is inevitably extracted along with the ore. It must be removed before the ore is smelted. The mineral content of deposits is derived in a variety of ways. The origin affects both their richness and the ease with which they can be worked.

Economic minerals like all other rocks originate in the igneous rocks which are the original solid stuff of the earth. Some are found in the igneous material itself, others in the sedimentary beds formed by deposition of the waste derived from its weathering and erosion. They are also found in the metamorphic rocks which are either igneous or sedimentary rocks radically transformed by subsequent heat or pressure. Yet others (long important to man in the case of salt) now recognized as a very much more numerous

* For explanation of this and other technical terms see Glossary.

8

and potentially very important class, are held in suspension or solution in sea water.

Some workable concentrations of minerals, that is 'ore bodies', were formed by concentrations within the still molten igneous material or 'magma' below the earth's surface, possibly by a process of segregation from the mass because of their higher specific gravity. From the cooling magma, gases and fluids penetrate the surrounding 'country rock' especially along joint or bedding planes or along the lines of weakness formed as the magma pushes its way upwards. The country rock around such a magmatic 'intrusion' is metamorphosed, the area most immediately and intensely affected being known as the 'metamorphic aureole'. In the aureole minerals separate out from the original gases and fluids according to their respective temperatures of condensation or solidification. Thus, around the granite 'bosses' which are often the surface expression of old igneous intrusions, are found a graded series of minerals – tin and wolfram on the inner edge and arsenic and copper further out. Similar high temperature solutions penetrating sedimentary rocks carried the load which was deposited in the carboniferous limestone of the Pennines or the lead and zinc of the famous Tri-State (Kansas–Missouri–Oklahoma) mineral district. When exposed at the surface, weathering and the slow movement of ground water produces an upper layer in which the mineral concentration is higher than in the interior of the ore body – a zone of 'secondary enrichment'. In some cases, however, it is the residue from such a process of weathering and leaching which is of economic import-ance. Some tropical laterites are low grade iron ores and, much more important, the weathering of silicate rocks under tropical conditions produces a residue of hydrated aluminium oxide – the bauxite which is the prime source of the world's present output of 10 million tons of aluminium a year. In other cases the materials removed from an outcrop by weathering and erosion are deposited elsewhere in commercially attractive concentrations. Heavy minerals soon settle out from the stream in which they are carried to form the valuable placer deposits of alluvial areas. The working of placers of gold, tin or diamonds characterized some of the most colourful episodes in mining history. Although gravel pumps and dredges have now replaced the more romantic pans of the pioneers, tin from such alluvial deposits still dominates in the output of Malaya, the leading producer. If leached from extensive deposits and then redeposited in concentrations or swept out into a lake or gulf of the sea which later dries out, other economically important mineral accumulations may occur. This is especially important in the case of the minerals which supply the chemical industry – as with the caliche (nitrates) of the Atacama desert, potash and anhydrates and common salt of Alsace, Stassfurt, Cheshire or Teesside and Cleveland.

Many minerals come from a number of different types of geological source. Iron ores are produced on a large scale from concentrations within magmas, as at Kiruna in Sweden, from replacement deposits in other rocks as probably at Bilbao, or from former lake or shallow sea deposits such as the hematite (Fe_2O_3) of Krivoi Rog in the U.S.S.R., now the most important iron producing district in the world, or the siderites (iron carbonates $FeCO_3$) or limonites (hydrated oxides of iron, $2Fe_2O_33H_2O$) such as those of the Jurassic ore fields of western Europe. Phosphates too come from an interesting range of sources. The mineral apatite ($3Ca_3P_2O_8CaF_2$), worked in basic igneous rocks, and most spectacularly in those of the Khibin mountains of the Kola peninsula in north-west U.S.S.R., is one source. Others are the accumulations of recent bird droppings which reached immense proportions on the Guano coast of the Pacific states of South America or the rock phosphates derived from guano impregnated waters of Nauru, Ocean and other islands in the west Pacific.

E.C.S.C. 43	▲ Western Advanced Industrial 180
	Communist Bloc 46
▲	Third World 15

▲ Third World includes all Latin America,
Middle East, Asia except Japan and China, Africa
except South Africa

▲ Includes Australia

World Iron Ore Production 1950

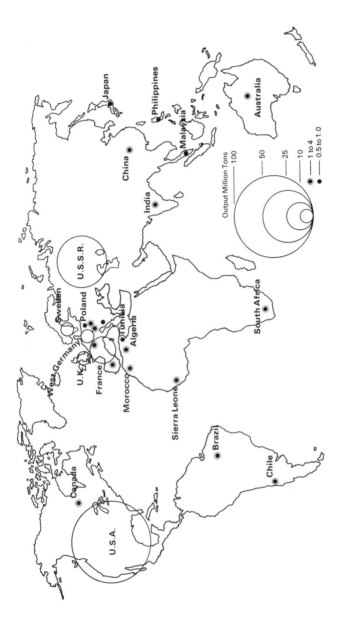

Figure 2   World Iron Ore Production 1950   Only producers of over 0.5 million tons are included

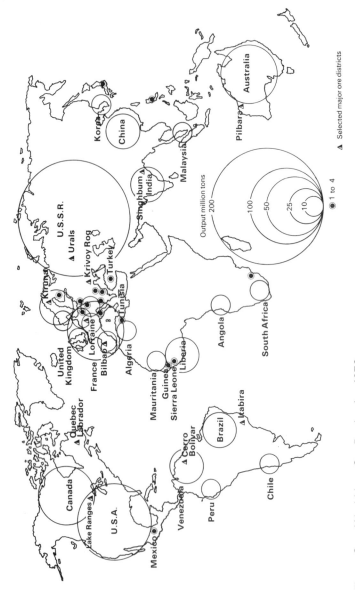

Figure 3　World Iron Ore Production 1970　Only producers of over 1 million tons are included

Fossil phosphate beds in Morocco and Tunisia provide the main supplies for western Europe, but along the Suffolk coast or in Cambridgeshire fields, at the base of the outcrop of the chalk marl, fossilized excrement known as 'coprolite' was worked extensively in the nineteenth century. The marks of these operations may still be found. Over the last ninety years another source of phosphorous for plants has been basic slag derived from the smelting of limey, high phosphorus Jurassic iron ore by the basic or Thomas process. In the last ten years fundamental changes in iron-making technology have caused the relative decline of this source. The table below indicates the great surge in world phosphate rock production, mainly for fertilizer production. The outstanding role of the U.S.A. (mainly from operations in Florida) and of the U.S.S.R. will be noted.

Table 1: World phosphate rock production (million metric tons)

	World	U.S.A.	U.S.S.R.	North Africa	Islands in Indian and Pacific Oceans
1934–8 average	11.3	3.5	2.0	3.6	1.3
1956–60 average	36.9	16.7	6.3	9.9	2.3
1969	77.6	36.0	15.3	13.6	3.9
1970	80.1	34.1	18.0	14.5	3.7

Source: W. T. Jones and *Mining Annual Review* 1970, 1971.

E.C.S.C. 70 ▲ Western Advanced Industrial 321

Communist Bloc 259

▲ Third World 167

▲ Third World includes all Latin America,
Middle East, Asia except Japan and China, Africa
except South Africa
▲ Includes Australia

World Iron Ore Production 1970

A broad pattern of world mineral distribution may be made out. Many of the metallic minerals, being closely associated with

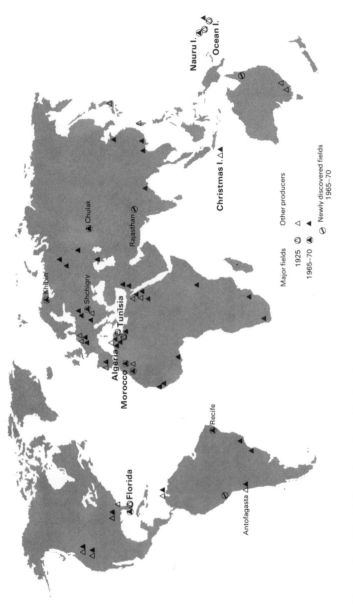

Figure 4   World Phosphate Rock Production

igneous activity, are found on the great blocks or shields which have been relatively stable elements in the geological structure of the world since Lower Paleozoic times, some 400 million years ago. Major examples are the Canadian Shield, that of Fenno-Scandia, at the surface in Scandinavia but also underlying much of eastern Europe and European Russia, and the Siberian Shield. Of rapidly growing importance in world mineral production are the now separated blocks which once formed the great southern mass known as Gondwanaland, in the plateau of Brazil, Africa and the western three quarters of Australia. Other great mineral concentrations are found in and around igneous intrusions frequently found in the core of the great mountain ranges which were shaped around the old shields in major, later periods of mountain building. Where the mountains have been deeply dissected these ore bodies are workable, as in the old mining areas scattered along the Hercynian mountains of Europe from Cornwall and Brittany through to the Harz Mountains and Bohemia. In the Urals and the mountains of Central Asia, and in the much more recent but still deeply eroded mountain belts or cordilleras such as those along the Pacific coast of the Americas or running from Yunnan through Malaysia to Indonesia, there are other major mineral provinces. In basins within the shields or mountain belts or in the more gently folded sediments along their edges are other minerals such as the massive bedded iron ores or the great oil, coal and gas fields. Man's mining

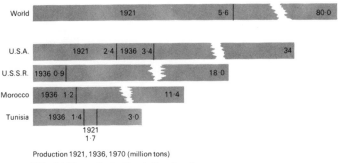

Production 1921, 1936, 1970 (million tons)

World Phosphate Rock Production

15

activity has picked out core areas within the shields or mountain belts as major foci of activity. Since the Second World War there has been a noticeable tendency to widen the range of search and development activity. This is related to an increasing knowledge of mineral resources and the sharp growth in demand.

The structure and mineral wealth of some parts of the world is clearly exposed. This is most obviously the case in heavily glaciated

1. *The Canadian Shield.*
In spite of difficult environmental conditions the Shield offers the advantage that frequently heavy glaciation has stripped away all superficial deposits and exposed the solid geology and structure.

shield country where geology and structure frequently stand out with extraordinary clarity in both surface and air views. Elsewhere the facts of solid geology are buried, sometimes below immense thicknesses of superficial deposits as in the great alluvial troughs of Mesopotamia, the Indo-Gangetic plain or Amazonia. Even in the case of the shields, mineral wealth may be hidden by thick glacial deposits and coniferous forest or, in the tropics, by deep weathering and rain forest. Fifteen years ago there was scant knowledge of the mineral richness of the Sahara. Australia's mineral wealth, which has since so spectacularly hit the headlines, was largely unknown because of a combination of adverse factors – size, small population, difficult climate, superficial deposits over large areas, lack of both demand and, perhaps fundamentally, of the 'atmosphere' of expectation of major new discoveries. In a 1953 survey the *Economist* recognized the Australian wealth in coal – though it decided the country was short of coking coal – put the iron ore resources at a very modest 100 million tons and acknowledged her wealth in lead, zinc and uranium. It concluded: 'The country is not particularly rich in other minerals but it can supply its own needs in copper.'[1] Yet in a comment whose wisdom has been proved by subsequent events, the same account noted that '. . . although most surface deposits in Australia have now been discovered and developed, no complete geological survey has been made and it is impossible to say how many minerals lie, as they are said to lie in the Sahara, below the barren surface of the Australian desert.' In many parts of those dreary deserts are the bonanzas and drilling locations of the late sixties and the as yet undiscovered or undisclosed deposits which will send speculative hearts fluttering in the seventies. On a world scale too growing demand has encouraged more thorough survey. At the same time the surveyor has been equipped with powerful new tools. The mineral deposits are a work of nature, the mining district the product of a human reaction to the opportunity they present. The state of geological knowledge limits the range of possible choices.

Mineral Resources

Mineral Survey

In the past the prospector was the chief agent in mineral discovery. A romantic, lone figure, sometimes skilled as a geologist or mineralogist, sometimes self-taught, he depended on surface finds and in most cases on locating the actual outcrop of the ore. Whereas now the big mining finance houses or the specialized mineral survey firms will search anywhere for ore deposits, the prospector was attracted to certain limited areas. One influence was the availability of sites where primitive peoples had their own small-scale operations, as in the Andes or the Great Basin of the western U.S.A. A large number of old gold workings made by the former inhabitants were found when white settlers arrived in Rhodesia. Early European efforts, though employing mining and pumping machinery, were largely confined to these sites. Another factor limiting mining to certain areas was the well established swarming of prospectors. Although the prospector was a lone wolf it seemed that the most likely place for a valuable new find was where minerals had already been found. This lead to a recognition of certain 'metallogenic provinces' or areas of richness in particular minerals, but as a result outlying, perhaps major occurrences, remained unknown.

Frequently major discoveries were made by chance, though in mining as in other things, chance favours only the minds which are prepared. The discovery of the silver ores of Potosi, Bolivia, in 1545, a discovery which was to create immense wealth and in the process cause the death of countless thousands of Indians in the mines, was said to have been made when a donkey tugged up the tree to which it was tethered. The discovery of the more prosaic iron ores was similarly frequently rather a haphazard business, even in nineteenth-century Britain or in the United States. Inland from Georgian Bay on Lake Huron the excavation of a cutting for the Canadian Pacific Railway in 1883 exposed the iron–copper–nickel sulphides of the Sudbury ore body, which since shortly after 1900 has been the world's outstanding source of nickel. Even into the mid twentieth century the occasional major discovery can be put almost in the category of chance. In the Second World War fears

18

about their long term fertility caused the dispatch of samples of red soil from the Claremont area of Jamaica for testing in London. It was proved that they were very rich bauxites. Within two years the war-booming American aluminium companies had acquired large interests.

Occasionally some extremely primitive early indicators of ore possibilities are still important. An account of prospecting techniques for bauxite in Guyana little over ten years ago provides an example: 'Since the bauxite is more resistant to weathering than the surrounding rocks, the deposits commonly stand up in slight relief. For this reason a favourite method of prospecting is to climb a high tree and look for "swells" in the tree-top level. This often marks hills of productive ore.'[2] Sometimes, even though elaborate survey techniques may be used, there is a chance element in that search for one mineral may result in the discovery of others. Crews drilling for possible oil fields in and after the Second World War revealed the eastward extension of the East Midlands coalfield much more effectively than ever before, and further north they discovered the potash deposits of Cleveland. In the fifties the Humble Oil Company found huge new sulphur deposits in the course of exploration off the U.S. Gulf Coast, and oil exploration also proved the Saskatchewan potash fields. The isolated prospector, the use of unsophisticated surface indicators, or the discovery of some other mineral than the one being sought, are all aspects of a haphazard search for minerals. A range of techniques has been developed in the course of the twentieth century to speed survey, to give it more precision and to permit discoveries at depths which would have been previously impossible without an extremely costly drilling programme, for which no reasonable justification could have been found.

A range of geophysical survey methods are now available. These trace out a variety of earth characteristics which may indicate possible mineral concentrations. A very important technique involves the magnetometer which measures variations in the intensity of the earth's magnetic field. Only iron gives a direct reading and some of the major successes of magnetometer surveys have been with iron. Examples are the spectacular new discoveries in the U.S.S.R. in the early 1930s, and the discovery in the early 1950s of the field at Marmora, north-west of Kingston, Ontario, where the ore body lies below more than 100 feet of limestone. Occurrences of iron pyrites are often associated with other metals and have distinctive magnetic characteristics, so that non-ferrous metals can also be outlined by magnetometer.

The gravimeter measures rock density. Again, this was used with great effect by Soviet geologists in their pioneer mineral surveys almost forty years ago. Electromagnetic survey makes use of the relationship between electricity and magnetism to trace the pattern of electrical conductors to a depth of a few hundred feet. It has proved effective in indicating sulphide ore bodies. Another group of techniques involve measurement of radioactivity, either through Geiger or scintillation counters or by means of gamma ray spectrometry. A major success from the use of the last was the 1970 discovery of very rich uranium deposits 130 miles east of Darwin in the Northern Territory, Australia.

Although each of these techniques has proved effective in the hands of ground parties, much greater speed is achieved by using aerial surveys. Aircraft have played a part in survey work since the early twenties; in the first place they permit the movement of ground teams into inaccessible areas with great speed, and, where the climate is difficult, especially in the northlands of Canada and the Soviet Union, this has very much extended the effective geological season. Aerial photography is now a major element in survey. Photogeology, as it is commonly called, began long before the war but its major achievements and wide elaboration have been concentrated in the last twenty years. At its least sophisticated it provides an overall view which the ground surveyor cannot gain, and so facilitates recognition of faults and fracture belts which have high mineral yielding possibilities. Stereoscopic analysis of overlapping pairs of photographs on scales between 1:20,000 and 1:50,000 yields impressive results in the hands of experts who can trace out such mineral indications as structure, rock type and vegetation, by distinguishing and recognizing the significance of as many as 200 different shades of grey in extreme cases. Colour photography, the comparison of infra red and optical photographic cover, or of high and low level photographs, open up still more possibilities. In spite of its small scale, satellite and spacecraft mapping gives a valuable new dimension. A geological map of a 830-mile-wide strip along the United States–Mexican border, prepared from spacecraft colour photographs, revealed information not readily apparent in conventional photography.[3]

Geochemical techniques involve the assessment of the mineral content of soil, rocks, water, or vegetation, from which it may be possible to trace concentrations back to a workable ore body. Geobotany makes use of vegetation as an indicator of possible mineral concentrations. The type or coloration of vegetation may indicate the presence of important structural features, define the limits of a rock formation or mark a focus of concentration of a particular element. The pipes of kimberlite which contain diamonds are found in ultrabasic rocks which, in the diamond region of Daldyn in the U.S.S.R., are indicated by a predominance of larch and alder. Work on the relationship between tectonics and mineral content has made great progress in recent years. Although well-established techniques continue to yield their crop of results, new devices are being introduced. In 1970, for instance, the Australian Scientific and Industrial Research Organization announced that it had successfully used a radioactive cored probe. This is lowered into a test hole and bombards the surrounding rock with neutrons. The records of the reflected wavelengths are analysed by computer to give a view of mineral prospects. This is the first device of its type in the non-communist world but the Soviet Union is believed to have a similar technique.

As a result of this continually elaborating range of survey techniques mineral exploration is evolving from '. . . a more or less haphazard scratching of the earth's surface to a systematic examination of the mineral potential of an area . . .'[4] However, the impressive results which modern ground or air survey may achieve are dependent on the geological facts, so that there is slight hope in some areas. The results are, even more obviously, conditional on an active pursuit of mineral knowledge. Differences in the pattern of intensity of mining, over the world map, reflect not only the differences in endowment but also in the vigour with which the

21

survey has been pursued. After long trailing behind the other advanced industrial countries, the U.S.S.R. in a few years in the period of the First Five Year Plan pushed through a most impressive mineral survey. By 1932, 5,956 scientific geological workers and helpers and at least another 100,000 seasonal helpers were in the field. The geologist H. H. Read noted that before the Second World War, when Britain spent £70,000 annually on its geological survey, the U.S.A. spent £1 million and the U.S.S.R. £20 million. As late as 1947 expert staff in British geological services overseas – in an empire which was then still intact – totalled fifty-eight, although independent mining concerns additionally employed their own geologists.[5] As late as 1970 it was estimated that of the 74 per cent of the area of Argentina which was believed to have mineral potential 68 per cent was still unprospected.[6] In the mid fifties, when scores or even hundreds of prospecting teams were active each season in Canada, the search for minerals lagged far behind in Australia. In 1956 the value of all Australian output was about $A430 million at present prices. By 1968 an amount approaching one tenth of this was being spent on prospecting, pegging and obtaining claims. Australia was arriving at the stage of mineral exploration fever which Canada had experienced over a decade earlier. The Mines Department in Western Australia handled about 3,000 claims in 1967: in the first three months of 1969 alone this figure was exceeded. This 'ethos' of exploration, the conviction that major discoveries can still be made, is of great importance. Australia produced its first nickel in 1967 and two years later the country was in the grip of a nickel fever. By this time a major nickel province was being proved in a great arc from 100 miles south to 300 miles north of Kalgoorlie. The value of nickel in one small part of this area, near Kambalda, is said to be worth more than all the gold produced by Kalgoorlie mines over three quarters of a century. Yet spectacular success though the discoveries have been, they have also chastened mineral surveyors, for prospectors, exploration engineers and geologists had all been active in this area for eighty years before this major base-metal deposit was discovered – in surface outcrops.

Ocean Mineral Resources

Sea water contains immense quantities of minerals but they are so widely diffused that only in cases where concentration is sufficient or where technical change or scarcity of other supplies forces new horizons on the mineral trades has it been much used. The most spectacular example of the first condition is with salt. For centuries common salt – with other minerals scattered in minute amounts through it – has been recovered in primitive salt pans round the coasts of the world. Salt extraction from sea water is more efficiently organized now and it is estimated that in 1968 35 million tons, or 29 per cent of world salt supplies, were derived from the sea.

Magnesium, as the lightest of the metals, is of vital importance in alloys for aircraft. Furthermore, it was extensively used in incendiary bombs. At the beginning of the Second World War Germany made almost 60 per cent of the world's small output of the metal. As war approached, Britain and the U.S.A. explored the possibilities of extracting magnesium from sea water, of which it constitutes 1 part in about 800, and the first British magnesium was made by this process in 1939. Although in the postwar period the magnesite of metamorphic rocks has remained an important source of magnesium, 61 per cent of 1968 world production was from sea water. U.S. consumption of magnesium was 103,000 tons in 1968 and the major sea water processing plant at Freeport, Texas, was being expanded to a capacity of 120,000 tons. World bromine supplies are derived to the extent of some 70 per cent from the oceans.

One cubic mile of sea water is estimated to contain almost 6 million tons of magnesium or twenty-eight years supply at the 1969 level. In this case extraction from what is for all practical purposes an inexhaustible source of supply is economical, but the existence of other metals in sea water is of no practical concern. As no technique exists for their economic extraction it cannot be claimed that they represent a resource. At the end of the Second World War it was said that it would cost £25 million to extract the £5 million worth of gold contained in a cubic mile of sea water. The profit line

23

for one material is very different from that of others, but as the exhaustion of land-based mineral resources proceeds and demand increases the likelihood is that more minerals will be commercially extractable from the sea.

Beach deposits are already important sources of a variety of minerals including iron and monazite – a source of 'rare earth' metals, of cerium and of the radioactive material thorium. Production of minerals from shallow water is common as with sand and gravel, oil, gas, sulphur and tin. In 1967 as much as 8 million tons of sand and gravel were worked in depths of up to 90 feet off the coasts of Britain. Off-shore operations provided 17 per cent of the world's oil by 1970 and the proportion is likely to be about double by 1980. Already oil prospecting has reached out to depths of 1,800 feet, though producing wells are all located in much shallower water. Effectively mineral extraction is still confined to the continental shelves, or depths not exceeding 600 feet, a very small part of the whole water surface. By the late 1960s the prospect of mining in the deeper seas beyond began to seem a practical possibility for the first time. As a result a completely new dimension in world mineral survey began to be recognized.

The oceans cover 140 million square miles or more than 70 per cent of the earth's surface. Their mean depth is over 12,000 feet. The traditional diver has been limited to operations not more than 250 feet deep, but, starting with the bathysphere in the 1930s, man has gradually devised craft which can penetrate to greater depths and which, moreover, are manoeuvrable. Yet only about 15 per cent of the ocean's floors lie at less than 6,000 feet. There are severe problems of mineral survey, not to mention those which will affect extraction, for this is 'a very hostile and cold environment, under pressure, often in negligible visability, subject to severe corrosion, with difficulties in breathing, communications and movement and often with turbulent or extreme weather conditions at the surface above'.[7] Exploration is a 'tedious combination of geophysical and geological measurement and assessment', extensive survey being followed by the process of proving which involves drags, grabs, use of divers or drilling. Operations are plagued with the difficulties of accurate positioning and fixing. Nevertheless the indications of mineral wealth at considerable depth are so numerous that active preparations are being made to develop mining.

By 1966–7, when the North Sea gas rush was in its most colourful phase, it was becoming clear that other, deeper off-shore zones offered equally rich prizes. Along the Atlantic seaboard of the United States petroleum deposits were proved in the section near the mid Atlantic states and off the shores of the Carolinas. Paralleling the shore is a long belt in which phosphorite (calcium phosphate) is concentrated and south-east of Charleston are large areas where the bed of the Atlantic is covered with manganese nodules. The potentiality of the ocean's yield of manganese nodules touched off the greatest excitement.

Manganese is a mineral of strategic importance which is extremely unevenly distributed over the land surface. It constitutes the major steel alloying element but few of the leading western advanced industrial nations have any home supplies. In 1968 the U.S.A. imported 2.3 million tons of manganese but produced only 11,000 tons of it: 95 per cent of her consumption is in steelmaking. In the following year the Communist bloc, producing only 28 per cent of the world's steel, mined half its manganese – the two biggest manganese producing areas in the world being Chiatura in Georgia and Nikopol in the Ukraine. The existence of manganese nodules on the sea bed was first indicated by the researches of the Challenger expedition almost a hundred years ago. Recognition that the nodules might be in sufficient concentration and at depths which now rendered them workable aroused great interest in the western world in the 1960s. The brownish-black nodules are commonly found at depths ranging from 5,000 to 18,000 feet but in some cases in much shallower waters. Their chemical composition varies widely but within the following ranges: manganese 5 per cent to 30 per cent, iron 5 per cent to 26 per cent, nickel 0.2 per cent to 1.8 per cent, copper 0.1 per cent to 1.6 per cent, cobalt 0.1 per cent to 1.0 per cent.[8] By 1966 Russian, Japanese, American and British firms were already interested in the possibilities of raising these nodules from the sea bed.

Technically there appear to be no insuperable obstacles. Specially designed deep sea dredges or hydraulic pumps suspended from a

25

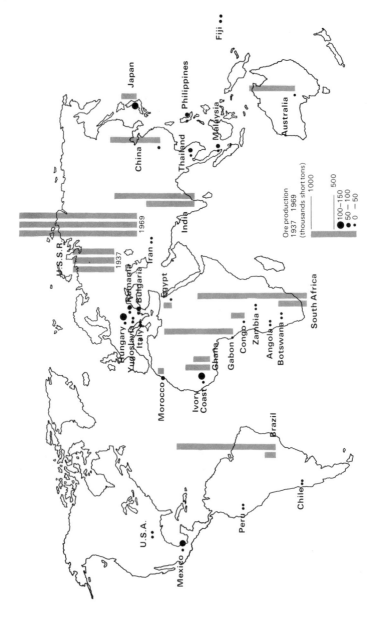

Figure 5   Manganese Ore Production 1937, 1969   United States Bureau of Mines

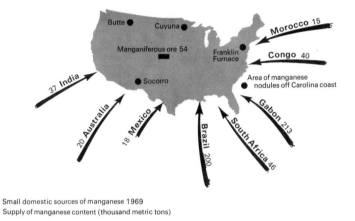

● Small domestic sources of manganese 1969
  Supply of manganese content (thousand metric tons)

Figure 6   U.S. Manganese Supply 1970

vessel could deliver to a land-based processing plant, whose design would have to be rather elaborate in order to deal with such a complex mix of metals. American estimates in 1966 put the capital cost of a dredging vessel designed to raise 10,000 tons of nodules a day from 13,000 feet at $6 million. Operating costs were reckoned at $2 per ton at depths of 3,000 feet but rising to as much as $5 at 16,000 feet. The processing plant would cost $60 million. These capital costs are comparable in total with those for a big new manganese mine in the developing world. However the latter has limited reserves as opposed to the practically unlimited long-term capacity of a dredge operation. Each operating unit would work a big area, estimates ranging up from 1,000 to 30,000 square miles. In some parts of the Pacific the nodule concentration reaches over 100 lb. per square yard and guesses at their availability over the whole ocean floor suggest that it contains enough nickel and copper to meet present world levels of consumption for 150,000 and 400,000 years respectively.

The repercussions of even a modest programme of deep sea mineral extraction are complex. One ship dredging 10,000 tons of nodules a day could supply annually more than the present U.S.

27

consumption of both manganese and nickel. World prices for these metals therefore would fall and some of the marginal producers would be eliminated from the lists. Indian manganese mines in particular would suffer.[9] Another intractable problem involves the international law of mineral extraction beyond territorial waters.

In 1958, when techniques for extracting minerals below 200 metres had not been elaborated – as opposed to limits well beyond that now – the Geneva Convention on the Continental Shelf adopted an inconclusive approach to the extension of national rights beyond that. It resolved that the exclusive rights of a coastal state extend '. . . to a depth of 200 metres or beyond that limit to where the depth of the superjacent waters admits of the exploitation of the natural resources'. Doubt remained concerning the situation where major islands owned by distant powers prevented the extension of national rights half way across each ocean and it was still not at all clear whether after all any mineral operator could not operate by right of capture in deep water anywhere in the world.

In 1967 the Maltese delegation to the United Nations suggested that the sea bed should be developed for peaceful purposes and in the interest of mankind as a whole, advocating that 'the net financial benefit derived from the sea and exploitation of the sea bed and of the ocean floor shall be used primarily to promote the development of poor countries'. In the summer of 1970 discussions were held both in Malta and in Geneva to develop a code for oceanic exploitation. At this time the U.S.A. proposed that all development at depths greater than 200 metres should be controlled by an international authority to which any nation or company exploiting resources there should pay royalties. The funds thus accumulated should be used to help both oceanographic research and the underdeveloped nations to exploit the resources of the sea.[10] It is unlikely that ocean bed mining will develop at the pace or with the glamour with which science fiction would surround it, for investment decisions must always be taken in full consideration of the costs of alternative, conventional land-based operations. Even so there seems little doubt that in the next decade survey of deep sea minerals will not only be widely extended but will pass over into the first stages of mineral extraction.

3. Mining – Some Aspects of its Economics

The Location of Mining

Mining is very unevenly distributed over the earth's surface.[1] In part, this is the result of the geological factors discussed in the previous chapter, in part the result of a whole range of other locational influences. Mineral deposits are fixed in location but they vary in mineral content, ease of extraction and proximity to market and therefore in their overall working costs. To meet demand a choice must be made between a variety of possible operations. When this is done a mineral 'deposit' or 'occurrence' makes the transition to a mine.

Even a preliminary consideration shows that various branches of economic activity are located according to very different principles. In an advanced economy the pattern of retailing is fairly directly related to the distribution of the population, with the proviso that in different parts of a country standards of provision and sales per head may vary. Routine office work is similarly widespread but head offices cluster in regional centres and still more noticeably in business or national capitals. Manufacturing industry has a bewildering range of locations. So-called 'footloose' industries, whose productions are of high value in proportion to their bulk or weight, like clothing, light engineering, or electronics, may be located quite widely within a national territory. Major assembly industries, which put together the finished products of other trades, are attracted to areas which have a complex of suppliers. The most spectacular example of this group is the motor industry. Bulk-reducing heavy trades need large tonnages of raw materials, usually of mineral origin, employ very great quantities of water in processing, and require big sites and facilities for the dispatch of their large output. Oil refining, steel and heavy chemicals are outstanding representatives of this class. Manufacturing concerns benefit from

29

large scales of operation and tend to cluster in order to gain agglomeration economies, that is shared facilities of various types.

Agricultural location is far removed from this. In the first place, the quality of the site becomes important. Whereas its location and its existence as a firm base for building is all that is important for shops, offices or factories, the farmer is concerned with the fertility of its soil, water conditions, aspect and the annual procession of the weather across it. Access to markets, either in the form of a direct consumer or a processing plant is the other vital consideration determining what shall be grown. The raw material supply aspect, important in manufacturing, especially with heavy trades, is a minor factor in farming. Here, therefore, there is a fundamental difference from the situation in manufacturing – in an already settled country, rather than look for a place to produce a particular crop the farmer has to take his location as given and to decide what it is best suited for. In mining the situation is different again.

A limited number of sites are available for the production of each mineral at any time, in other words, those deposits known to possess reserves workable in the current state of technology. As with farming, resources are fixed in location, though a choice can be made between them, and other factors of production, men and capital, must be moved to them. In some cases the number of occurrences of a mineral is so few that choice for development is very much simplified – the historic confinement of a very large share of the non-communist world's production of nickel and asbestos to parts of eastern Canada is an outstanding example. Usually the range of choice is much wider. In addition to the 'pure' economics of mining each major industrial country may follow a partly politically motivated policy. As with some basic manufacturing industries, it will develop certain of its mineral resources so as to provide a check on the freedom of others to exploit it with extortionate prices or to control part of its essential military supplies. Lack of full knowledge about mineral resources, the historical legacy of existing mines and the political factor mean that at no time does the world pattern of mining reflect what would result if the mining department of a world confederation could decide which mines to operate on strictly economic criteria. A simplified, rational view of the location factors for mining does however provide a good starting point.

Given deposits of equal richness and ease of extraction, those near to major markets will be opened first. However, immediately one runs into the snag of assuming too much about the scale of working and the definition of the market. As in manufacturing, mining overheads mount less rapidly than scale of operation, so that, other things being equal, the bigger the operation the lower the unit cost. However, even in a world with no impediments to free trade, costs of carriage on low grade minerals may be so high that even apart from the wish to avoid strike risks, 'acts of God' and so on, it may be desirable to open two or more mines of less than optimum scale rather than one, bigger mine. The value of the product is vital here. Diamonds, platinum or nickel will stand transport well, gravel or bricks will not. The economic advantages of big operations in the gemstone diamond pits around Kimberley are not cancelled out by transport and distribution costs to the multitude of jewellers' shops in the world's cities. At the other extreme the operating cost advantages of huge pits working sand and gravel are quickly nullified by haulage costs, so that each great city or area of major industrial construction work must find its own near-by sources of constructional raw materials. Broadly speaking the lower the price for the mineral, the more burdensome the costs of carriage and the less easy it is to supply demand from one source. It is true that if highly efficient bulk rail or water transport and high capacity/low cost handling facilities are employed then operations may be fairly well concentrated even when the mineral is of low value. A notable British example is the concentration of the brick industry in the Oxford clay vale especially around Bedford and Peterborough. Although brickmaking survives elsewhere, this area has important advantages in rail and road transport facilities, access to coal and excellent local clay. It has extended its markets widely even though, as the chairman of the London Brick Company revealed in 1968, at points over 120 miles from their works – which made a record of 224 million bricks in November 1967 – over half the delivered cost was made

up of haulage charges. Concentration in brickmaking is sympto-
matic of an important change in mining location. Since the Industrial Revolution began there has been a trend to more large-scale
mining with ever wider markets dependent on improved transport
facilities and marketing organization. Over the last 10 years this
process has been extended so rapidly as to bring in for the first time
on a world scale what the mining correspondent of *The Times* has
referred to as the 'new philosophy of large scale, low grade
mining'.[2]

So far deposits of equal richness and ease of working have been
assumed, but if differences are introduced here then new combinations of locational considerations are possible. Small or low grade
deposits may be workable near to the market when further away it
is essential to work richer or larger ones. The Marmora ore body
in Southern Ontario is small and of relatively low grade – 17
million tons of 37 per cent iron ore when production began in 1955.
However, it is well located for rail shipments to the works at
Hamilton and Lackawanna (Buffalo). The iron content of the
Northampton Sand ore bed is 27 to 34 per cent iron but that of the
Frodingham ore of north Lincolnshire only 20.4 to 23.2 per cent.
In both areas there are important steelworks on the orefield – Corby
and the three Scunthorpe works respectively. In 1969 no Frodingham ore left north Lincolnshire. In the same year 4.35 million tons
of Northampton Sand ore were dispatched from the district of
production. Of this 3.34 million tons was railed to north Lincolnshire but 0.68 million tons travelled as far as the north-east coast.
As in so many other cases the situation is not so simple as at first
sight. Chemical variations in the ore make certain mixes desirable
and this involves the movement of Northampton Sand ore to
Lincolnshire, but not a reverse movement.

Other complications involve costs of production. There are some
variations in these in the case considered above. Thus, while
Frodingham ore is worked below a thicker overburden this is
mostly heavy clay as opposed to the thick bed of limestone as well
as of clay in the quarries of the Corby area. Moreover as the ore
immediately east of Scunthorpe lies below poor wind-blown sands
there is not the same cost involved in restoring the surface after

working as is the case in Northamptonshire. Environmental factors push up the costs of metal mining in Swedish Lapland or in the high Andes. Many of the problems of copper working at Cerro de Pasco stem from its height of 14,400 feet in the Peruvian section of the cordillera. In the twenties for instance, shortly after the mine had converted from its traditional silver production to an emphasis on copper, its physical difficulties made the company wholly dependent for unskilled labour on local Indians. Geologists, engineers, foremen and managers had to be brought in from outside at high salaries and each year had to be permitted one or two visits to a sea level community to recuperate.

The situation of gold in Canada is one in which both access and conditions of work vary. The major focus of Canadian gold mining is in the camps of western Ontario. Conditions here are harsh enough but they are less severe than in the north-west as, for instance, at Yellowknife, on the north shore of Great Slave Lake. The railway has recently been pushed north from the Peace River country to the southern shore of the lake, but until this was done Yellowknife was a remote area. In the mid fifties it was reckoned that deposits with as little as 0.15 ounces of gold per ton of country rock could be worked profitably in Ontario but a minimum for profitable working in the North-west Territories was 0·45 ounces per ton.[3]

Cleveland iron ore mining provides an example in which all three factors, quality of ore, conditions of working and costs of transport all became progressively less easy as development proceeded. In 1850, when the Cleveland main seam was first worked on a large scale on the outliers on the northern edge of the Cleveland hills, open-pit working was adopted. From these operations there was a short, gravity-powered journey down to the main Middlesbrough to Redcar railway and the works below. The centre of mining gradually moved further south or south east so that transport costs to the ironworks increased. Within a few years adits and shafts had replaced open pits and costs of underground haulage, lifting and pumping were increased. In the eastern basins of the hills the pits were especially deep although this was compensated by their large reserves. As mining moved south so it was found that the original main seam was split by a shale parting which gradually increased in width so that eventually the two iron-stone beds had to be worked separately. Finally, as working moved away from the outcrop, so the iron content of the ore fell away and the proportion of undesirable constituents increased. With costs increasing in a range of ways, so also technical change in the metallurgical operations – the switch from iron to steel, and then, partly for institutional reasons, the delay in switching from acid to basic steelmaking processes – made this phosphoric ore less attractive so that foreign hematite progressively ousted it from the local markets. After the Second World War ore production was subsidized in order to maintain employment, but subsidies were withdrawn in 1958, and six years later the last mine closed.

However, it must be reaffirmed that scale economies may permit successful working of a low grade, relatively intractable ore far from the market. Perhaps the outstanding example of this, more fully examined later, is the replacement of the small, fairly high grade copper operations of western Europe or eastern North America by the low grade, disseminated copper deposits of the western States and a little later of Chile and central Africa where operations were on a much bigger scale. Copper also illustrates another factor – 'externalities' or the cost advantages or dis-advantages accruing to a particular plant from its local conditions. In spite of its large chimney the big smelter in the Butte district of Montana damaged vegetation and reduced amenity, but in the pioneer west this was of little significance as compared with the depredations of smelters near to the smaller eastern operations where the wrath of woodsmen and cultivators and later of con-servationists could more easily be aroused. The smelter at Duck-town in eastern Tennessee was notorious in this respect. Similarly, although they have many other advantages, the tin smelters of Bootle are more hedged about by planning restrictions than the small but expanding refining capacity of the Bolivian Andes.

In summary, to say in the time honoured phrase 'minerals are where you find them' seems a confession of woeful ignorance of the complexity of the factors which shape the geography of mining. Alternatively it may perhaps be looked upon as a wise aphorism, a recognition that no simple explanation is possible and that one may as well put aside most preconceptions and start with the pattern as it exists. However, the mining concern, having studied demand prospects, may decide that new capacity of a particular mineral is necessary. It will follow a number of distinct stages from that point until the mine is inaugurated.

2. *Shaft sinking.*
Sinking the shaft at the new Creighton Mine, Canada. In completing the 7,137 foot deep and 21 foot diameter shaft the amount of rock removed was sufficient to fill a train thirty-five miles long.

35

Stages in the development of a mine

Before a mine begins production the project passes through a number of distinct phases, extending over as much as five years and involving survey, evaluation, construction and running-in. The capital investment and the risks increase as the concern passes from one stage to the next. Prospecting and the initial discoveries are the work either of the traditional lone prospector or, and increasingly, the result of the use of the elaborate survey methods already considered. Costs are naturally at their lowest at this stage although, even so, one recent Australian estimate for big mine development put them in the range $A25,000 to $A150,000 (or £11,500 to £70,000) and some $A40,000 million was being spent on prospecting annually in Australia in the late 1960s.[4] The next stage involves the evaluation of the ore body and of the complex of factors which will affect its working, leading through to a decision either to work it or to abandon or postpone development. Whitmore suggested that for a big mine this stage might cost anything from £230,000 to over £900,000. Almost twenty years ago the American Mining and Smelting Company gave up its options on lead–zinc deposits in northern Nigeria because of underground flooding. It had already spent almost $2 million on the investigation of them. The ore body must be outlined by drilling, so that its shape and size can be determined and the grade of its ores appraised. In addition, mining, milling, smelting and refining costs are projected, and it is necessary to put a figure to the cost implications of scale of operation and the infrastructure which will be needed – and which in an under-developed country or region may involve transport and power facilities, a township with all its services and so on. Technical, marketing, accountancy and tax specialists will be involved. The political situation and stability of the country in which the deposits are located will be important and a growing consideration. The long term profitability of the project may be assessed by the net cost flow. This flow will be negative at the start as the mine is built and run in but will become positive, that is profits will be made, as time passes. From the profits deductions have to be made to service the investment, for taxation and for depreciation, which in the case of mining involves the eventual writing off of the whole enterprise as the ore body is worked out. As, over the long term, money values are likely to fall, the basis of assessment of the viability of a project – the vital decision to turn an ore body into a mine – is the discounted cash flow (D.C.F.), both positive and negative elements of the cash flow being discounted to the same date.

There follows the development stage which, in Whitmore's words, is 'the least speculative of all, approaching the respectability of investment'. However it may be a very costly investment and even in some circumstances a risky one as construction times for a mine lengthen, and the project is therefore subject to the danger of production beginning at a time whose price levels can only be roughly estimated.

Methods of Working

Stage three of mine development involves commitment to one of a number of possible methods of ore extraction, usually either underground or open-pit working. In exceptional cases extraction by other means is possible.

The working of minerals as fluids involves great economy of equipment and still more of labour. Unfortunately it is limited to extraction of liquids, gases or minerals which can be easily dissolved or held in suspension. Advances in drilling techniques in the postwar period have included angled bores or even lateral ones at considerable depths and these have widened the physical circumstances in which fluid mining may be possible. The earliest and still the most important examples are oil and natural gas. By the 1880s brine was being dissolved and pumped from depths of as much as 1,000 feet on Teesside, and by the early 1960s only 1 per cent of British salt output was mined. Early in the present century the methods pioneered by H. Frasch were first applied to the extraction of sulphur from the salt domes along the Gulf Coast of the U.S.A. The Frasch method involves the application of superheated steam and compressed air through a central pipe, the dissolved material being forced up an outer pipe to settling tanks on the surface. In both the Saskatchewan potash fields, opened up in the late fifties and the early sixties, and the potash field of Cleveland, now being developed a quarter of a century after its discovery, some operations use solution mining, others shaft mining. At Belle Plaine, west of Regina, Saskatchewan, potash – and necessarily also other soluble salts – are dissolved from their solid impurities underground. Surface operations consist largely of evaporating plant. Solution mining permits the working of lower grade deposits at greater depths than conventional methods partly because the proportion of the total deposit extracted is higher.

3. *Drilling iron ore preparatory to blasting at Kiruna, northern Sweden.*

Underground mines still play a very large role in world mineral production. Vein deposits are not usually suitable for open-pit operations and even disseminated deposits may be more suitably worked by an underground mine if the overburden is thick. A variety of techniques is involved. From the main shaft the levels may follow the veins or, where the mineral is widespread through the mass, the 'room and pillar' technique may be used. Again 'stoping' may be employed, ore being systematically worked out at various levels. In 'caving', the mineral falls, extraction taking place from a lower level.

Underground mining involves high development expenditure – even in 1957 shaft sinking cost anything from £20 to £100 and the driving of levels £3 and upwards a foot.[5] As compared with open pits, operations are still rather labour intensive, the ore must be lifted from the base of the shaft and, as working proceeds, the amount of underground haulage goes up. Pumping costs may be high, a notable example being the Zambian copper mines. In September 1970 this brought tragedy to Mufulira mine, the biggest producer of the Roan Selection Trust group, a 'cave-in' there of 1 million metric tons of water, sand and mud causing loss of eighty-nine lives and the interruption of full output for anything up to a year. On the other hand, underground mines are less subject to the vagaries of the weather. When the nickel ores of Sudbury, Ontario, were opened eighty years ago underground operations were preferred to open-pit mining partly to permit year round operations. On the other hand, when the copper–nickel deposits of Norilsk east of the Yenisei, U.S.S.R., were developed in the thirties, open pits provided most of the output in spite of even more severe climatic conditions.

There are more unforeseens in underground than in surface mining. As depth of operation increases not only will costs mount but the type of ore may change, requiring new methods of extraction or processing. Worse still the ore may run out and the operation and an isolated community are then faced with extinction. There are, however, some compensating advantages. As opposed to the mass, indiscriminating extraction process in an open pit, the quality of the product can be varied according to economic circumstances.

Lodes vary in width and richness, so that, when prices are high, grades may be lowered, thinner veins and poorer ore being worked. In times of depression the opposite policy will be followed. In the old Cornish mines rich portions of ore were left here and there to furnish a steady supply when other parts of the mine were unproductive, an operation known as 'picking out the eyes of the mine'. In short, underground mining is 'selective mining'.

The ultimate stage of underground mining involves remote-controlled, continuous operations saving both on the capital and direct operating account. So far this ideal has eluded the mining engineer. Esterhazy mine, 150 miles east of Regina, the biggest potash operation in the world, represents an important step towards it. The ore body is covered with impervious strata so that there is no water problem and the ore is firm. Instead of blasting and drilling, there are continuous mining machines, driven electrically and mounted on caterpillar tracks. Two automatic cutters enable each machine to work a thirteen-foot face producing five tons of ore a minute. This is fed through the machine to a conveyor belt, on to a shuttle car and to the base of the shaft. Bevercotes colliery in the East Midlands coalfield is another highly sophisticated, remote-controlled operation but ran foul of management/union controversy which reduced its impact. It was then closed as a result of geological difficulties but re-opened in summer 1971.

Open pits are non-selective or 'mass mining' operations. Large, even if low grade deposits are required and success depends upon highly efficient mechanization in the removal of overburden, extraction of ore and in the removal of material from the pit. Open pits now work large, fairly rich deposits down to 800 or 1,000 feet, but drainage and haulage costs increase with depth, so that at this stage conversion to shaft operations may occur. In populated areas open pits involve a greater outlay than underground mines because of the difficulty and expense of acquiring and restoring land, but this is not so in remote areas or where regulations are lax. The desolation caused in the Appalachians or in Queensland by open-pit coal extraction is the result.

Open-pit mining requires efficient overburden stripping. This involves huge 'walking draglines' with a very big load capacity and

4. *Walking Dragline.*
This large dragline is operating in the open-pit coal operations of Utah
Development Corporation in Queensland.

a boom long enough to swing the waste well outside the area to be
worked. By the end of the sixties the biggest unit in the world,
operating in the Indiana coalfield, had a lifting load of 385 tons
and a 250 foot boom.[6] The ore is then extracted, in favourable
circumstances, such as those in the Caribbean bauxite fields, simply
by power shovel, more usually by explosives followed by shovelling.

5. *Open-pit blasting at Phalaborwa.*
In this Transvaal copper operation some 80,000 tons of copper concentrate are
annually processed at the plant near to the open cast workings.

The second critical element in open-pit operating is the quick and cheap removal of the ore. Haulage costs altogether total 40 to 60 per cent of total production costs in typical large open-pit operations.[7] In the past much ore was removed by rail, the track being continually realigned as working proceeded. In recent years huge dumper trucks have become more popular. At the beginning of the sixties open-pit dumper trucks were commonly of 22 to 40 tons capacity but by the end of the decade ranged up to 85 to 120 tons. It has proved difficult to design and to maintain a sufficiently big

6. *High capacity dumper trucks.*
A 100-ton Dart truck being loaded with iron ore at Mount Tom Price in the Pilbara District of Western Australia.

power unit for these trucks. Tyre wear has been another problem. With some of the larger trucks, outlay for tyres ranges up to 40 per cent of operating costs and 20 per cent is a normal figure. Even though operating costs have fallen there is scope for even more economy. Estimates in 1968 suggested that if, in place of the present material flowline (shovel → truck → semi-permanently located crusher in the pit → conveyor → mill), the truck was eliminated so that shovels emptied directly into mobile crushers, costs could be cut by as much as one third from an estimated 18.3 cents per ton moved to 12.2 cents.[8]

7. *Large open-pit operations, Marquesado Mine, Spain.*
Bull-dozing into the crusher and transport by conveyor belt at the Marquesado Mine.

Open pits have been making ground at the expense of underground mining. In operations of smaller capacity, of 0.5 million tons annual output or less, underground mining predominates. In bigger units open pit mining comes into its own. In 1968 fifty-two mines were brought into operation in the non-communist world;

a little under half were underground but of these only six were in the 1 million tons plus annual capacity range: twenty-one of the open pits were of this size. Projections made then suggested that underground mine output would go up by 21 per cent and that for open pits by 28 per cent over four years.

There are limits to the prospects of open-pit mining. For the single pit, depth and haulage costs, water problems and so on, are limiting factors. The ore grade may fall off with depth and as this happens the amount of waste increases. Inflation pushes up equipment costs at a rate which cannot be contained by even the most careful attention to cost paring. On the other hand there is considerable scope for all open pits to move nearer the standards of efficiency of the best. Figures of 'cost per work output' – a figure which takes into account tonnage handled, vertical and horizontal distances over which it was moved and friction – have been shown to vary widely for fourteen American open pits. The median value was 0.0213 cents but the extremes were 0.0086 cents and 0.0454 cents.

Scale Economies

As with manufacturing so in both shaft and open-pit operations, impressive cost advantages can be won from big operations, and the scale is rising rapidly. Overheads do not rise in proportion to capacity, especially in an underdeveloped area where investment covers a broad spectrum which in a developed area would already be part of the infrastructure. The development cost of eleven major projects in uninhabited areas of Australia in the sixties was about $800 million. Infrastructure costs amounted to $515 million of this total. Labour costs per ton are lower in large-scale operations. World mining is still dominated numerically by relatively small mines but in response to these conditions big units are becoming more and more prominent. In 1967 984 mines in the non-communist world, a little under one seventh of the total, produced 96 per cent of the output of all metal ores and other minerals except coal. In the following year twenty-seven out of fifty-two new operations were of 1 million tons a year or over, a size category with only 5.3 per cent of the existing units.

Table 2: Number of mines of various sizes for minerals other than coal in the non-communist world 1967 and 1969 (thousand tons/year)

	Under 150,000	150,000–500,000	500,000–1,000,000	1,000,000–3,000,000	Over 3,000,000	Total
1967 approx. 6000		403	197	253	131	6984
1969 approx. 6000		427	198	256	159	7040

Source: Mining Annual Review

Scale economies may enable production from lower grade deposits or operations disadvantaged in some other way. U.S. mines working ores of 1 per cent metal or less have been able to meet the competition of 3 per cent copper ores from Chilean or Zambian mines where the flow from the pit was frequently smaller and where labour was not so efficiently used, though nominally cheap.

Financing Mineral Development

In the past it was common to start mining with a relatively small output, making progressive extensions to new, higher levels when demand and profits justified. Now, with world demand so insistent and the advantages of scale so clear, it is common to plan for a big operation from the start. Costs have been rising, so that a big new mine may cost $100 million to $300 million or more to develop with no return for perhaps five years and a very uncertain price situation when production begins so far ahead. Early in 1970 an agreement was signed between the Southern Peru Copper Corporation, an American concern, and the Peruvian government for the opening of the Cuajone deposit at a cost of $350 million. Development work began in April 1970 and is expected to take between six and seven years. At the end of March 1970 the price for copper on the London Metal Exchange was £755 a ton; by October it was £450. When Cuajone production begins prices may have risen again, though the general view seems to be there will be some surpluses in copper until the late seventies. There is obviously extreme uncertainty about the return on such a capital-intensive project. This makes the problem of financing a serious one, and increases the desirability of some tied outlet.

The days of the small prospector who went into a small operation and built it up into a great mining concern are perhaps not over, but the more general line of development is now quite different. Commonly a successful prospector will sell his discovery to a mining finance house which undertakes evaluation and development. A small group of internationally renowned firms provide much of the capital for big mine projects and have the reputation to secure most of the rest. Some of them at first operated in the South African gold fields but have widened their interest into other minerals and other countries. Major developments over the last twenty years have been financed by one or a consortium of these firms. In the underdeveloped world they may work along with loans from international agencies. In recent years this method of financing has been supplemented by fixed-interest borrowing secured against a prospective long-term cash flow which, in turn, is ensured by long term pre-emptive sales contracts. Such an arrangement is of especial interest to mineral deficient nations not traditionally involved in international mining. In the sixties Japan has emerged as far and away the most important member of this group.

As a secure outlet and a source of some of the finance, Japan is concerned in many great mine projects throughout the world – iron ore very widely, coal mining in eastern Australia, South Africa and British Columbia, copper in Iran, Indonesia, Bougainville island in New Guinea, Papua and many more.[10] In 1969–70 when nickel supplies were tight and prices were rocketing, some of the major expansion projects were linked with Japanese needs. The French company Le Nickel is planning a new ore processing plant estimated to cost over $100 million on the north coast of New Caledonia.

Japan obtains about 90 per cent of its nickel ore from this island, is helping to finance the plant, and has negotiated a contract for long-term ferro-nickel supplies.

Mineral Costs and Prices

The costs of producing ore are affected by a host of factors. On balance these are less easy to predict or to control than those affecting operating costs in manufacturing for there are more unknowns, more hostages to fortune. The richness of the ore and its chemical composition, physical characteristics of the body and of the mining district itself, and its location in relation to smelter and market are obviously important. Other factors are scale of operation, charges and the tax policy both in the producing country and the firm's home government.

Mine, smelter, and probably also township, railway and power stations and other items of infrastructure represent a very large commitment in capital and therefore in fixed costs. Yet demand changes according to the state of the world economy. Mineral prices have varied within wider limits than those for manufactured goods.

It is true that where control rests with a few major groups a fair degree of price stability has sometimes been maintained. Aluminium, and historically – though certainly not recently – nickel, are

Table 3: Export Prices for Non-ferrous Base Metals and Manufactured Goods 1958–69 (1963 = 100)

	1958	1960	1961	1962	1964	1965	1966	1967	1968	1969
Non-ferrous										
Base Metals[1]	92	106	102	100	119	135	156	142	150	168
Manufacturers[2]*	97	98	99	92	101	103	106	107	106	109

[1] Price indices [2] Unit value indices
* World leading producers

Source: U.N. Statistical Yearbook 1969, 'World Exports'.

impressive examples. On the other hand, demand for tin has grown only slowly and there are a large number of mines including many small producers and widely varying production costs in a number of underdeveloped countries, so that tin prices have been extremely variable. Copper prices too have followed a most uncertain course. It is the basic material of the world's electrical industries and even fear of shortage sends prices disproportionately high. This encourages more exploration and mine development and prices react strongly. In less than a year to early 1971 the copper price on the London Metal Exchange fell from almost £750 a ton to little over £400.

The very epitome of price unreliability is wolfram, the main source of tungsten, about 95 per cent of which is used by the steel industry. Most of the non-communist world producers are small, many of them have high costs. Variations in demand push prices fiercely up or down, the number of producing mines varying even more widely. Many mines were at work at the time of the Korean War but closed afterwards. It is said that in 1955 there were over 700 wolfram mines in the United States but by the end of 1958 only two were still at work. In two weeks in 1959 world prices for wolfram rose from 100 to 160 shillings a unit.

Sometimes mineral price changes are catastrophic, as with the fall in base metals which marked the end of the steady postwar expansion in world output in the late 1950s. The shareholders who suffer from such slumps have been mostly located in the industrialized world. The Union Minière du Haut Katanga was a source of income for many Belgian families, Indonesian tin and oil revenues comforted many a Dutch home and British investors gained from mining profits on a world-wide scale. Finally one recalls the photographs of American annual company meetings

Table 4: Base metal prices 1955–7 (year incl.). £ per ton.

	1955	1957
Copper	436	162
Lead	126	71
Zinc	105	61
Tin	887	730

similar to those which Anaconda and Kennecott must hold – solid middle-class citizens and particularly the serious, keen and rather disturbing faces of the elderly ladies. But even in the last decade of colonialism the price oscillations in the world's metal markets hit the underdeveloped world hardest, and now that partnerships, national marketing organizations or outright nationalization have replaced the old expatriate company, the underdeveloped countries will be still more exposed to the ups and downs of mineral prices.

Minerals and Economic Development

The landscapes of the coal and iron ore districts of Britain and some of her base-metal districts, notably west Cornwall and the lead dales of the northern Pennines, indicate eloquently enough the major role which mining can play in economic development. Their present Development Area status indicates what happens when mining is conducted as a robber economy, with the expenditure of the minimum necessary for operations in the mineral district, little economic diversification and profits remitted elsewhere. In Britain much of the wealth made from mining and metallurgy went to build up big estates in the early days, and later to shareholders who frequently lived in the softer, southern areas of Britain. The case has its parallels on the bigger scale of the developing world.

The underdeveloped world – common usage 'the developing world' is a euphemism which hides both the enormity of the gap between poor and rich countries and the fact that in most cases it is getting wider – plays an important role in world mineral supply, though the balance of production still lies heavily with the developed world (Table 5).

Western mineral investments caused substantial economic development in the colonial world. The mine, railway and port brought in or trained new skills, increased purchasing power and extended market areas for the farmer and trader as well as the mining company. Even so development was patchy, leading to the characteristic 'island' pattern of development, areas of modernity and growth set within a sea of scarcely changed tribalism and

49

Table 5: National Wealth and Mineral Production late 1960s

	G.N.P. per capita 1966 ($)	Iron ore	Bauxite (1968)	Copper*	Zinc*	Lead*	Tin*	Manganese	Sulphur*
Leading Advanced Industrial Economies and nearby surplus mineral producers									
U.S.A.	$3,520	90.0	1.7	1.4	0.5	0.5			9.5
Canada	$2,240	32.4		0.5	1.2	0.3			3.5
Japan	$ 860	2.0		0.1	0.3	0.1			—
Australia	$1,840	38.5	4.9	0.1	0.5	0.4	0.007	0.8	
United Kingdom	$1,620	14.8					—		
W. Germany	$1,700	7.4			0.1	—			1.6
France	$1,730	56.0	2.7		—	—			0.5
Italy	$1,030	1.1	0.2		0.1	—			
Sweden	$2,700	30.7			—	0.1			
Spain	$ 640	6.1		0.1	0.1	0.1	0.005		
Total (median income $1,715)		279.0	9.5	2.2	2.8	1.5	0.012	0.8	15.1
U.S.S.R.	($ 890)	186.0	5.0	0.9	0.6	0.4	0.027	7.0	1.6
Developing economies									
Chile	$ 510	11.6		0.7					
Mexico	$ 470	3.9		0.1	0.2	0.2			0.1
Peru	$ 320	9.0		0.2	0.4	0.2		1.5	1.7
Brazil	$ 240	29.0	0.3						
Malaysia	$ 280	5.5	0.8				0.072		
Thailand	$ 130	0.5					0.020		
Indonesia	$ 100		0.9				0.016	—	
India	$ 90	28.5	0.9					1.8	
Zambia	$ 180			0.7	0.1	—		—	
Congo	$ 60			0.4	0.1		0.007	0.3	
Total (median income $ 230)		88.0	2.9	2.1	0.8	0.4	0.115	3.6	1.8
World total output		701	46	5.8	5.4	3.3	0.176	19	28.4

Note: * Metal or other final product content. Rest are ore tonnages (million tons).
 — Marks small output.

Source: International Bank and *Mining Annual Review.*

primitive farming. The mining camps of the altiplano had little impact on the drudgery of the Andean Indians. It is only a few years since primitive tribesmen were shooting poisoned arrows at Venezulan oil field workers, and the railway linking Livingstone to the mining agglomerations of Broken Hill and the Copper Belt merely carved a swath of improved land through the Zambian savanna. The financial contribution of mining to state revenues in the early days was minimal.

Mexican oil exploitation provides an admittedly extreme example. Until after the First World War Mexico was one of the world's leading oil producers. Before 1917 neither royalties nor taxes were paid by the oil companies, which even operated their own armies. Even after this, for some time the royalty was only 5 per cent. Subsequently Mexican production fell away sharply, mainly as a result of restrictive legislation, heavy taxes and even government support for strikes in the oil areas. Foreign oil companies were expropriated in 1938.

The underdeveloped mineral producers sometimes complain that the minerals they produce are exported in their low value form, further processing being undertaken elsewhere. They therefore press for more smelting. A number of considerations are involved in any rational resolution of the problem of smelter location. The more intractable the ore, that is, the more technical problems processing presents, the more desirable it is to locate the smelter in an advanced industrial country. Lead, tin, copper, zinc and aluminium, in that order, present increasingly difficult smelting problems (see Table 6).

In the mid fifties, the twilight years of colonization, although two thirds of the world's bauxite supplies came from underdeveloped countries they had less than 1 per cent of the aluminium capacity. On the other hand, more than half the copper from the mines in the underdeveloped countries was smelted there, and two thirds of their lead was smelted at the source of supply. There are other factors to be considered. The size of the ore body is important, a big operation being necessary to justify a local smelter. Such a unit will be expensive to build and its construction and operation may require

Mineral Resources

Table 6: Smelters in the Underdeveloped World 1956

	Lead	Tin	Copper	Zinc	Aluminium
Africa	4	4	6	3	–
Middle America	7	1	7	1	–
South America	7	3	9	3	3
South and S.E. Asia	8	11	6	5	10
Total	26	19	28	12	13

Source: U.N. Non-ferrous Metals in Underdeveloped Countries, 1956.

highly paid, imported skills. A significant local demand for the metal, nearby power supplies and a favourable fiscal policy may secure a local smelting operation much earlier than would otherwise be the case. Finally world location is important. If smelters in an advanced economy are easily accessible, this, and also their probably greater scale economies, may inhibit local smelting. Mexican mineral processing is said to have been affected by the closeness of the U.S.A.[11]

The initiative in the relationship between mining company and underdeveloped country has recently shifted to the latter. Thus in the mid sixties Nigerian tin producers were squeezed by a royalty scale related to the metal's price. Inflation pushed up not only prices and therefore the royalty, but also costs. Nigerian tin mining companies therefore received only one third of the price increases. National control over mining concerns advanced rapidly throughout Africa and Latin America in the form of 'Congolization', 'Zambianization', 'Chileanization' of mining and, more recently still, the new Peruvian mining law. For the mining country increased control and revenue resulted. In the short term there can be no doubt that even the big mining groups are at their mercy. In the longer run it is possible that this will cause a major shift in the map of world mining, with exploration and development concentrating on the areas which are politically more stable, such as Australia or Canada. The underdeveloped countries undoubtedly have good reasons to feel that they have been shamefully treated in the past, but it has proved easy to move too far, or too rapidly, to a position in which the mineral producing economy is dominant. Consideration of the postwar history of mining in Burma and Indonesia is salutary.

Before the war Burma was a regionally important oil producer, from the Yenangyuang fields in the middle Irrawaddy valley. Bawdwin mine in the Shan States was probably the world's largest lead mine, and also an important source of zinc and silver, and the Mawchi̇ mine in Karan state the world's first ranking tungsten mine and the third producer of tin. In 1938–40 minerals made up 38 per cent of Burmese foreign trade. The economy was greatly disturbed by the Japanese invasion. In 1947 Burma left the Commonwealth and for years afterwards was ravaged by insurrection. In 1962 a new military government began a vigorous nationalization programme. It is known that Burma is rich in minerals although two thirds of the territory has still not been effectively surveyed. Copper, iron, coal and nickel alone to the value of £700 million have been proved, but the country is in fact self sufficient in oil and little more. Minerals had fallen to no more than 2 per cent of total export values by 1948 and by the late 1960s still represented no more than 4 per cent. Meanwhile the country suffers extreme poverty, the estimated per capita income in 1964 being less than 4 per cent of the U.K. level.

Indonesia, with four times Burma's population, and vast resources, is also poor and underdeveloped. Apart from its oil and tin it has important reserves of bauxite, iron, nickel and manganese. Efficient Dutch paternalism made it a model of the old idea of a colony, dependent and unbalanced, and, in return, as a recent Australian account put it, Indonesia lived in 'a state of prosperous subsistence'.[12] After the war the Dutch were expelled and economic mismanagement led to inflation, a breakdown in trade and the nationalization of foreign enterprise. The Sukarno regime's confrontation with Malaya cut off the immediate outlet for one quarter of the country's exports, including its tin ore. In 1966 Sukarno was effectively replaced by General Suharto, and economic stabilization policies were begun. Early the following year a foreign capital investment law was introduced to encourage foreign capital along with local participation. Mining is to be a joint government/company venture. Together with agricultural improvement, mineral development is the chief target of development planning to 1973.

There seems little doubt that present policy will encourage much new foreign investment, for, especially as far as Japan is concerned, Indonesia is a highly attractive mineral supplier. Even so the twenty years of withdrawal, partly in chaos, have left a backlog of development work. In 1968 the Indonesian minister of state for economy, finance and industry wrote frankly, though unfortunately without a precise definition of some of his terms, about the mineral economy. 'The popular world view is that Indonesia is a treasure house of natural resources. In all frankness, nobody knows exactly the wealth there is, in and on the ground and in the seas around Indonesia. Only 5 per cent of the country is on the geological maps, 40 per cent on the geographic and under 1 per cent on the oceanographic maps. In short, Indonesia is waiting for the west to bring its advanced technology, experience, capital and entrepreneurial spirit.'[13]

Mining companies have noticeably liberalized their approach over the last quarter century, an inevitable adjustment to the changing context of their operations. An important statement of the range of responsibilities of a mining concern was given to the Royal School of Mines in 1957 by Sir Ronald Prain, then Chairman of Rhodesian Selection Trust (now Roan Selection Trust).[14] Sir Ronald recognized that major mining operations can have a large effect on the economic and social conditions of the region and its relation to the nation, on the state of the nation itself and, beyond that, affect world trade in primary produce and all its dangerous oscillations. He then spelled out seven responsibilities of the mining company, responsibilities which are by no means easily reconciled. The 'first and obvious' responsibility was to the owners, that is shareholders, and stockholders. The second and third were to employees and the host government, a fourth was to the community in the district where the mine was located, and here Sir Ronald pointed out how easy it was to upset the national economic and social balance by establishing anomalously affluent nuclei in a poor nation. It would be desirable that the effect of these islands of wealth and western technology should radiate outwards

54

to the whole economy but in practice they polarize economic growth, denuding the rural areas of their young workers, their more talented and enterprising members, and attracting new industries. Sir Ronald noted that, whether 'as a moral responsibility' or 'as enlightened self interest', some mining companies had already chosen to spread the effects of their operations by boosting power developments, or by investing in transport or in agricultural improvement, as some of the copper belt companies were doing. The next responsibility was to customers and then to competitors, consultations with whom could secure more stable trading conditions for all. Finally there was 'a responsibility towards the future', the recognition that with a wasting asset something must be done to ensure the future prospects of all with interests in the present wealth from the operation – through conservation, research and the pursuit of geological exploration. Recognition of this range of responsibilities led Sir Ronald to enunciate a mining philosophy, which though clearly capable of specific interpretation, was too generalized and middle-of-the-road to satisfy newly emerging independent nations for long. 'If these interpretations of an industry's responsibilities are accepted, we are bound to arrive at the conclusion that modern management philosophy dictates a way which is designed to be a compromise between the more ruthless type of private enterprise and the more sentimental type of public interest.'

Since that statement of 1957 there has been a rapid evolution of the relationships between mining companies and host governments. In 1969, speaking to the Ninth Commonwealth Mining and Metallurgical Congress, Sir Val Duncan of Rio Tinto Zinc outlined the new, still more liberal policies which the mining houses have espoused in reaction to the pressures exerted on them. His statement was echoed in the paper 'The Philosophy of an International Mining Company' given by M. I. Freeman, also of Rio Tinto Zinc, to the Peruvian Mining Engineers Convention at the end of 1969.[15] The mining company must regard itself as a co-partner with the country in which it works, must train its citizens and move them as rapidly as possible into positions of management, work towards a higher degree of local autonomy and give the local population the chance to acquire financial interest in the concern. This is a far cry from the autocracy with which mining operations were conducted fifty years ago or less, but the underdeveloped world remains subject to all the vagaries of primary product price uncertainties and therefore to difficulties in terms of trade. To provide them with a greater measure of justice it is necessary that a liberalizing influence should spread much more widely than merely through the corridors of the head offices of the great mining houses.

4. Mineral Transport

Mineral deposits are localized so that development involves a choice between areas of potential production. Among other things this choice is conditioned by the availability and cost of transport. In a developed country only land transport may be involved to link mine and smelter but, as the sphere of mineral marketing has widened, so increasingly both land and sea transport are employed, with the need for bulk transfers between them. In the case of precious metals transport costs are a significant, but rarely dominant, consideration; with base metals low transfer charges are a prerequisite of successful operation.

Until the nineteenth century exploitation of mines distant from markets or from the coast or from inland navigable water was costly. The pattern of production of such a low grade mineral as coal was closely shaped by this consideration. The canal system selectively widened the marketing radius of inland mineral districts, and by the mid-nineteenth century railway construction enabled coalfields in the English midlands to market for the first time in London at prices competitive with those from the north-east. Similarly improved access by rail encouraged expansion in British base-metal districts before foreign fields, also made accessible by shipping, port and railway construction, began to undercut them. In most cases, where there was a pre-existing scatter of mines, direct railway connection was not achieved by all, so that road hauls to a loading point were still common, as in the lead dales of the northern Pennines or in the copper and tin districts of the southwest. At the same time small ports, now long forgotten as important foci of commerce, imported coal for the mines and shipped out their produce to the smelters. A century ago Hayle, on St Ives Bay, received vessels of 200 tons, with coal, timber and general cargo for the mining district around Redruth, to which it was connected by rail. It had regular steamship sailings to Bristol. Portreath, whose capacious docks have in postwar years handled only minor coal traffic, then had a tram railway to the famous Gwennap mines and forwarded their copper to Swansea. St Agnes, though accessible only to 100 tonners, handled coal, lime and slate. Portmadoc, at the terminus of the Ffestiniog Railway, shipped large tonnages of Welsh slate and, although Parys copper operations were already a shadow of their late eighteenth-century glory, a small steamer still sailed weekly to Liverpool from Amlwch harbour, cut out of the slate rock, and said to be capable of accommodating thirty vessels of 200 tons each.

In the days of pack-horse transport and small sailing ships, Spanish America was world-renowned for its precious metals. One of the most suggestive indications of the importance of cheap bulk transport is the switch of some of its mines to large-scale, base-metal production on the arrival of the railway. This happened at Cerro de Pasco, Peru. For 250 years, from the discovery of the district's mineral wealth to the 1890s, emphasis was on silver, but by the end of the period silver output was falling away. An eighty-three-mile line was built from Cerro de Pasco to Oroya, providing a connection with Lima. A shorter line was built northwards to local coal deposits, and by 1907 Cerro de Pasco was being transformed into an important copper producer. Later lead and zinc production was added. For the whole of Andean America the opening of the Panama canal in 1914 was a very important development. The ports of the previously remote Pacific seaboard became accessible on a shorter, less stormy route to Europe, and, still more important, to the eastern seaboard of the U.S.A. This affected all base metals, especially copper prospects, but one of the most spectacular changes involved iron ore.

As late as 1911 the notice of Chilean minerals in the *Encyclopaedia Britannica* observed: 'Iron mining has never been developed in Chile, although extensive deposits are said to exist.' When this was written work on the Panama Canal was well advanced, and in 1913, anticipating the major change in space relationships which would result, the Bethlehem Steel Corporation of eastern Pennsylvania

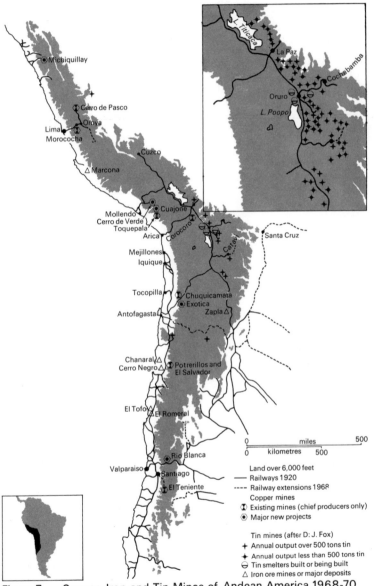

Figure 7   Copper, Iron and Tin Mines of Andean America 1968-70

bought the rights to the undeveloped El Tofo iron ore deposits near La Serena in central Chile from the Schneider interests of France. Bethlehem Steel's rapid rise to complete pre-eminence in the steel business of the mid-Atlantic region of the United States was associated with its initiative in opening the El Tofo mines.

The undeveloped nature of the interior of Africa at the start of the twentieth century delayed mineral exploitation. In the first decade tin mining was begun on the Bauchi plateau in northern Nigeria. Operations were limited to panning, and the ore was carried by porters in 60 lb. sacks as much as 190 miles to the river Benue and from there down river to the coast. The overall freight cost to the shipping point was reckoned as much as £12 5s. a ton. By 1910 the Lagos–Zungeru–Kano line ran some way to the west of the Bauchi plateau, and by the end of 1914 a 143-mile branch line had been built from Zaria to Bukuru across the tin mining area. As railway access was improved, so it became economic to extend mining and improve the methods of operation.

More recent, successful development of giant operations in low grade ores has depended on increases in the efficiency and scale of rail handling facilities, wagon-to-ship transfer and ore movement in specially designed vessels. Conversely only very large-scale mining can justify the huge outlay for integrated development projects – the mine and its ancillary production facilities, railway and port. The opening of the mineral wealth of Mauritania is a case in point.

Mauritania is the former French territory, 419,000 square miles in area, or more than four times as big as the United Kingdom, on the Atlantic coast of the Sahara. As might be expected from its position, it is a poor country having a population of under one million, largely nomadic and occupying the land at an average density of no more than two per square mile (as compared with 570 in the United Kingdom). Mineral exploitation is now recognized as the road to an improved standard of living. By 1966, even though still under $70 million, mineral exports represented 95 per cent of the total, a higher percentage than in any African country except Libya, then in the middle of its oil boom.

It was recognized that the black mountain, Kedia d'Idjil, rising

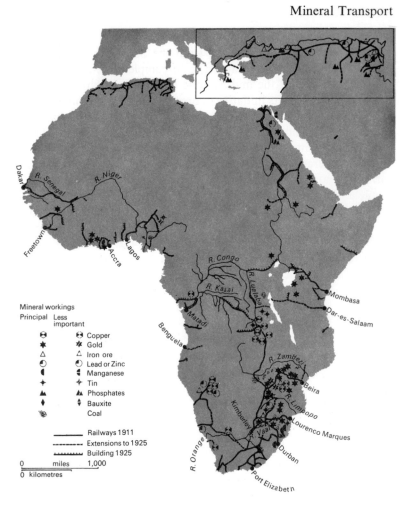

Figure 8   African Minerals and Transport 1925

out of the yellow, infertile plain near the military outpost of Fort
Gourard in the bulge made by Mauritania into Spanish Sahara,
was largely made up of iron – a mass twenty miles long containing
an estimated 120 million tons ore of 65 per cent iron content. Long
distance and desert separated it from a barren coast. To open the

61

ore body lavish investment in infrastructure was essential. Indeed, as M. Armard, Chairman of the African Industrial Bureau, summed it up, it was necessary to mine five to ten million tons of ore or none at all. By 1963 a railway had been built following the shortest route within Mauritania, along the international boundary, to Port Etienne on the Cape Blanc peninsula. Here a dock for 100,000-ton ore carriers has been built. Mine and port give Mauritania its two first enclaves of modernity, or, as the *Observer's* correspondent put it, '. . . two islands of modern ingenuity and comfort, linked by a 400-mile single track desert railway'.[1]

By the end of the sixties the harbour was being deepened to fifty-two feet, and rail movement had increased to two trains each way daily, those to the coast hauling 14,000 or 10,000 tons each, and on the return almost completely empty. In 1970 shipments were 9.2 million tons of which almost one fifth came to Britain. In spite of its mineral wealth and revenue, the Mauritanian national income per head in 1966 was only about $125.

Further mining promises a rise in this low level and a spread of the development effect. Big copper deposits were discovered at Akjoujt less than 200 miles north-east of the coastal capital of

8. *Akjoujt Copper Mine, Mauritania, 1969.*
A view of the mining area from Copper Hill.

Nouakchott. In 1968 the International Finance Corporation, the European Investment Bank (E.I.B.) – the E.E.C.'s investment arm – and others, committed themselves to large shares through the Societé Minière de Mauritania (Somima) which is investing at least $60 million. About 29,000 metric tons of copper will be produced annually. Mauritania will gain from taxes, duties and its share as

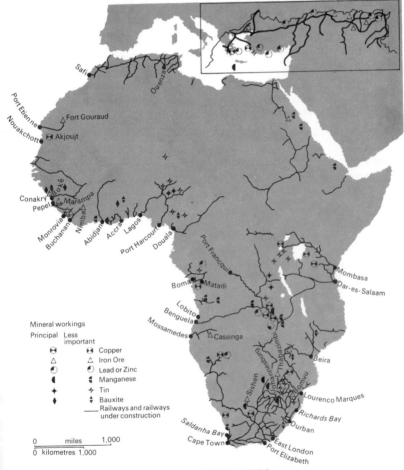

Figure 9   Mines and Railways of Africa 1970

an investor in Somima. A crusher, treatment plant for the refractory ore, power station, mine township and a 70 km. water pipeline will widen the impact. As the distance to the coast is less and copper concentrate will stand transport better, a tarred road replaces the railway of the iron-ore development. At Nouakchott the port will be improved, but, instead of bulk loading plant, vessels anchored off shore will be supplied by lighter. At present Nouakchott merely imports 33,000 tons a year, but by 1975 will

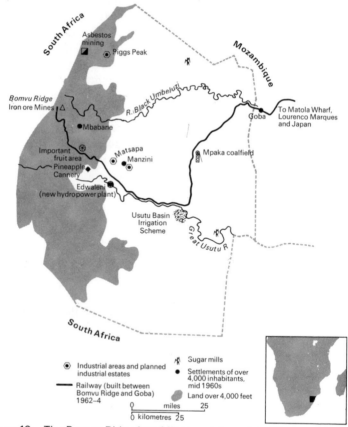

Figure 10   The Bomvu Ridge Iron Mines and Associated Economic Developments

handle 75,000 tons general cargo, and the copper concentrate.

Similar mine–railway–port complexes of development are common around the African coast now. The Liberian development Bomi Ridge to Monrovia was one of the earliest. In the Republic of Guinea the Conakry project has been important. Following marketing agreements with Japan, Swaziland opened the Bomvu Ridge mines and, after construction of a new 137-mile line connecting with the existing route to Lourenço Marques, shipments began in 1964. In this case, as opposed to that of Mauritania, the railway could have a wider development effect. It was routed so as to help fruit- and sugar-growing firms and to take out the produce of the new paper mill at Usutu. A hydro-electric plant was built to serve the mine and a small colliery opened to supply locomotive coal. The growing size of African mineral shipments and their long haul nature has made it desirable to use bigger ships and so forced the steady improvement of mineral ports. By 1967 dredging had opened the Sierre Leone iron ore shipping port of Pepel to 40,000-ton vessels. Early in 1968 long-term contracts signed with Japan involved the deepening of the ore dock to take 80,000 to 100,000 tonners. Improved ship handling, ore stocking and faster loading was also involved. Pepel loaded its first 100,000 ton carrier in 1969.

In the fifties Canadian mineral development was similarly multifacetted but in this case often under difficult, sub-arctic conditions. The iron ores of the trough stretching north from the area around Wabush Lake on the Quebec–Labrador boundary to Ungava Bay were discovered in the 1890s. The 1894 Report of the Geological Survey of Canada noted:

> The most important geological information obtained is the discovery of a great and hitherto unknown area of Cambrian rocks, extending north-northwest from North Latitude 53 degrees to beyond the west side of Ungava Bay. Their chief economic value is due to the immense amount of bedded iron ore . . . Thick beds of fine ore . . . were met with in many places and the amount seen runs up into millions of tons. Owing to their distance from the seaboard, these ores at present are of little value but the time may come when they will add greatly to the wealth of the country.

The first serious prospecting and the first claims issued by the Quebec Ministry of Mines date from 1919, but no ore was shipped until 1954. The key to development was the 360-mile railway from Seven Islands on the St Lawrence, which accounted for half the original cost of $250 million. The terrain was difficult and the climate harsh but construction of the railroad was speeded by the remarkable expedient of flying in parts of the track to airstrips along the route, construction taking place not from one end or the other only, but in a series of sections.

9. *Heavy mineral train, Lapland.*
The iron ore route from Kiruna to Narvik, the most northerly railway in Europe.

The railway has made it economic to open other iron ore deposits and there has been more railway construction inland from the estuary as well so that in 1968 the Canadian Eastern Area, as defined by the American Iron Ore Association, and almost wholly centred on the Labrador Trough, accounted for just one quarter the shipments of ore and agglomerates produced in the U.S.A. and Canada. The area is of course quite unsuited to all-round economic development. However in some cases railway building, even in the inhospitable environment of the Canadian Shield, has had a snowball effect. Mining at Chibougamau in central Quebec 300 miles north of Montreal is a case in point.

Copper ores, with some gold too, had been known for many years in this area, but it was the completion of a road from the Lake St John district in 1950 which brought extended survey and the realization that the deposits were much bigger than previously believed. By autumn 1957 three copper mines were at work. Concentrates had to be carried by lorry 159 miles south-east to Felicien and then transferred to the Canadian National Railway for carriage to the smelter at Noranda to the south-west of Chibougamau. In November 1957 a direct Chibougamau–Noranda rail link was opened. This cut out the road/rail trans-shipment and reduced the distance from 740 miles to 295. The estimated saving to the copper producers was $2 a ton. In response to the new opportunity two more mines were prepared; immediately also two newsprint companies began to cut pulpwood along the railway route.

Canada was the pacemaker in world mineral development in the 1950s. In 1942 a record output for all classes of minerals – fuels, industrial and metallic minerals – was reached of $567 million. By 1955 the figure was $1,795 million. Growth has continued, but less spectacularly; 1968 output was $4,700 million and that of 1970 $5,800. Although there were some discoveries in the southern more accessible and populous parts of the country, such as Marmora iron ore, Albertan oil and natural gas and Saskatchewan potash, much of the growth depended on roads or railways to inaccessible northern mineral fields. In the sixties the role of international leader in new mineral projects passed to Australia. Expansion in this case has largely depended on access to desert areas.

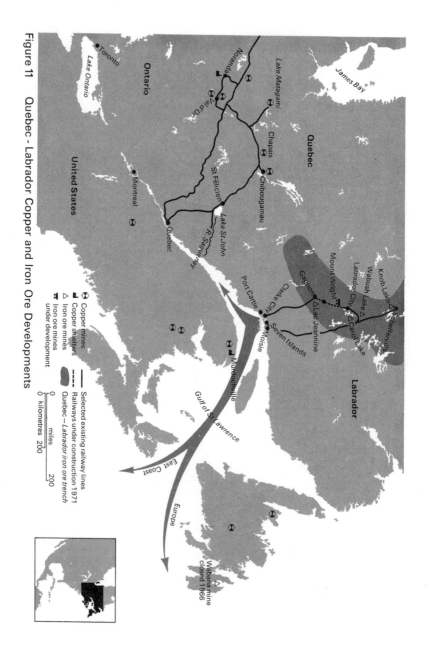

Figure 11   Quebec - Labrador Copper and Iron Ore Developments

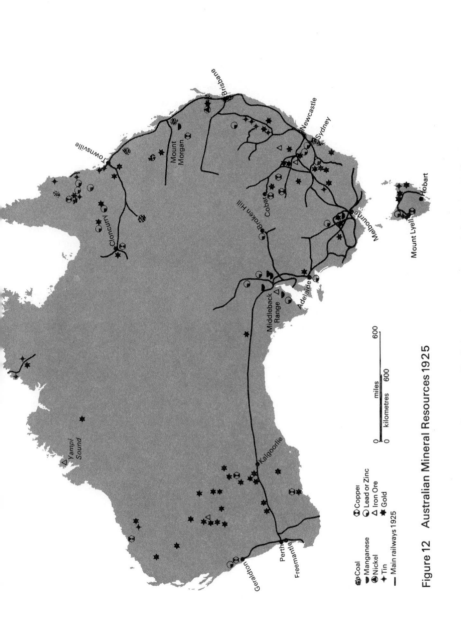

Figure 12   Australian Mineral Resources 1925

Coal
Manganese
Nickel
Tin
Copper
Lead or Zinc
Iron Ore
Gold
Main railways 1925

0   miles   600
0   kilometres   600

Townsville
Cloncurry
Mount Morgan
Brisbane
Newcastle
Sydney
Cobar
Broken Hill
Middleback Range
Adelaide
Melbourne
Mount Lyell
Hobart
Kalgoorlie
Geraldton
Perth
Freemantle
Yampi Sound

Mineral Resources

Before the Second World War most Australian mineral production came from deposits opened in the nineteenth century. This was the case with such famous mining camps as Broken Hill (lead, zinc, silver) in the west of New South Wales and the copper mines of Mount Morgan, Queensland, and Mount Lyell in western Tasmania. The major exception was the lead, zinc, silver and eventually also big copper production of Mount Isa in the Selwyn Range of western Queensland, and the neighbouring smaller mining camps to which the railway from Townsville was completed in the second decade of the twentieth century. The Newcastle coal basin, the Middleback iron ore range of South Australia and scattered small gold mining operations made up most of the rest of the Australian mineral industry. In the year 1939–40 44 per cent of the value of Australian mineral output was gold. A large number of surface deposits were known but the mineral geology of the continent was not. It must be noted also that after the Second World War the population was still less than 8 million, and most of the interior is either desert or poor pasture and the east–west extent of the state is as great as from the west of England to the Caspian Sea.

Postwar years added the discovery of the uranium of Rum Jungle, of Mary Kathleen and Radium Hill. Weipa bauxite in the Cape York Peninsula was proved in the mid fifties. In 1956 the value of mineral production was equal to $A462 million, and exports were $A165 million. By 1968–9 exports were $A759 million. In 1965–66 minerals provided 11.9 per cent of Australian export earnings; in 1969–70 24.2 per cent. Long term contracts concluded by early 1970 will involve an export business of $A1959 million by 1976–7. As new contracts are negotiated the figure will in fact be much above this. Increase in value has been accompanied by a shift to lower grade minerals, notably to iron ore, bauxite and coal. New products have been brought into the list and there has been a large increase in the number of producing districts.

In 1956 coal was the only one of the three low grade minerals exported. Bauxite mining had not yet begun and export of iron ore was prohibited. Coal exports were valued at 1.2 per cent of the figure for all minerals. By 1976–7 the share of mineral export value

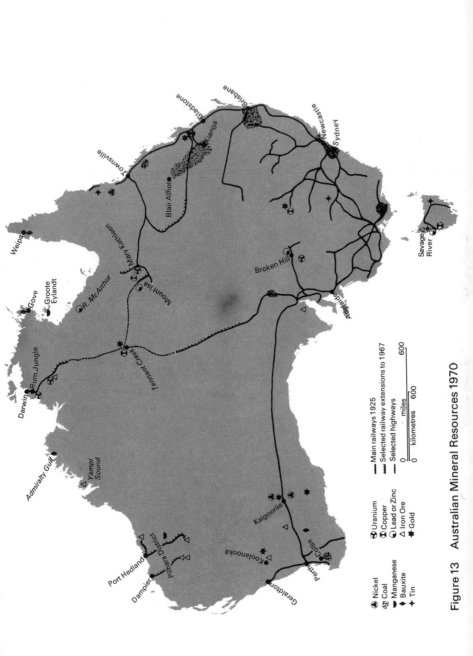

Figure 13   Australian Mineral Resources 1970

Nickel   Uranium
Coal   Copper
Manganese   Lead or Zinc
Bauxite   Iron Ore
Tin   Gold

Main railways 1925
Selected railway extensions to 1967
Selected highways

0   kilometres   600
0   miles   600

Weipa
Gove
Groote Eylandt
Darwin
Rum Jungle
Admiralty Gulf
Yampi Sound
Port Hedland
Pilbara District
Dampier
Koolanooka
Geraldton
Perth
Collie
Kalgoorlie
Tennant Creek
Mary Kathleen
R. McArthur
Mount Isa
Broken Hill
Adelaide
Blair Athol
Mt Morgan
Gladstone
Townsville
Brisbane
Newcastle
Sydney
Savage River

made up of coal, bauxite and iron ore will be about 63 per cent of the total. Except for coal, concentrated near to the populated centres of the south-east or the ports of the Queensland coast, much of the increased output will come from areas remote from both people and ports, and where environmental problems are great. Under these circumstances scales of production must be big, ports suitable for super ore carriers, and heavy duty railways to link the mines to them must be provided. The most impressive growth has been in Western Australia.

Table 7: Australian Mineral Exports 1956, 1968–9, and 1976–7 (m $A)

	1956	1968–9	1976–7 (projected)
Silver	12	20	12
Gold	18	23	8
Lead	70	86	152
Zinc	18	50	77
Copper	18	50	156
Coal	2	117	300
Tungsten	7	6	10
Bauxite	*	93	368
Iron	*	193	568
Tin	*	10	13
Manganese	*	12	24
Salt	*	1	14
Oil and Gas	*	27	42
All other	21	71	215
Total	166	759	1959

* Included in all other.

Sources: Various and The Economist, 28 February 1970.

As little as ten years ago maps of Western Australia showed little mineral development in the north except for a few goldfields and the iron ore workings in Yampi Sound. Now a host of new mineral workings have appeared with iron ore mining the most prominent. As with Labrador–Ungava the iron ore wealth of the area has long been sketchily known, first accounts of some of the iron ranges now being opened going back a century.* Western Australian mineral working was however long confined to gold, both in the north and south of the state, and to the Collie coalfield in the extreme south-west. As with the country as a whole it is important to keep the scale in mind. Western Australia is over three quarters

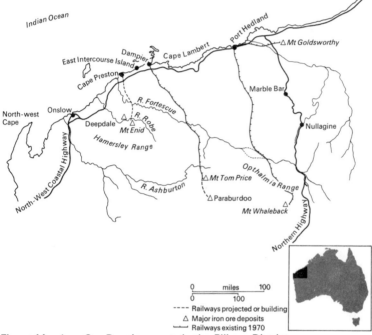

Figure 14   Iron Ore Developments in the Pilbara District

the size of India and almost twenty times the area of England, yet its population, even now, is only about that of Liverpool. Its overall density is similar to that of the Amazon Basin. More than half the population lives in the Perth–Fremantle industrial area and the other main populated areas are in the long established grain and dairy belts of the south-west. The biggest iron ore discoveries have been made in the ranges rising up out of the semi-desert of the

* If the reader consults the economic maps of Harmsworth Atlas of half a century ago he will see that the iron wealth of the ranges was already indicated.

Pilbara region of the far north. Rainfall totals in this area are low but there is extreme variability and hence occasional flooding. Summer temperatures reach as high as 120° F. or more, insects are a nuisance and there is little or no natural comfort in what has been described as '... an uninhabited, barren land, scarred with red rock outcrops and covered with scrub grass and twisted eucalyptus trees'. In the early 1960s Pilbara's population was less than 5,000.

Growth of Japanese demand for iron ore, and the lifting of prohibitions on Australian ore exports in 1960 was followed by state encouragement to exploration and mining. As a result very big reserves were proved. Exploitation has involved construction of new ore terminals at Port Hedland and Dampier, the first linked to the Mount Newman iron ore body by a 265-mile railway line and Dampier to Mount Tom Price by a 180-mile line. New townships have been built. High wages have to be paid to attract a work force and to retain it. By 1970 a very mixed population had been assembled at Mount Newman – under 600 workers but forty-seven different nationalities. Whereas the national average wage for men is $78 a week, in Pilbara even unskilled men earn $100. These high costs are balanced by the large scale of the operations. Hamersley Iron Proprietary will produce 17.5 million tons of ore and 2.5 million tons of high grade iron pellets a year from Mount Tom Price. By late 1970 some $A330 million or £155 million had been invested, of which over two thirds was for infrastructure. The most important item has been transport. Only $A60 million was spent on the mine but $A63 million on the railway, $A19 million for rolling stock and $A83 million on the development of port facilities at Dampier. Additional but unknown amounts have been spent on the construction of super ore carriers in Japan to deliver the ore to Japanese steel plants. Already by 1969 Dampier could handle 100,000-ton vessels and work was actively going on to make it suitable for even bigger ones. The railroad represents a fixed investment whose efficiency can be improved still more by increasing the motive power, but with the port the rapid increase in vessel size involves further massive outlays. This may include the partial writing off of the investment in Dampier, though that port will be used by smaller, if still by general standards very large, vessels.

74

10. *Dampier iron ore port, Western Australia.*
The newly developed port of Dampier is able to ship on the ore carried 180 miles from Mount Tom Price in 100,000 ton carriers.

Cape Lambert a little way to the east is reckoned the best location for a terminal to take 250,000 tonners.

The port and railroad development in the Pilbara district is being paralleled elsewhere in the world. In iron ore there are similar developments in many African states, in Brazil, India and elsewhere. On a smaller scale – because the tonnage involved in the metallurgical operations is smaller – the pattern is mirrored in the bauxite and alumina operations of Queensland, Guyana and Jamaica, the development of the Boké bauxite ores in Guinea and with other base metals, notably copper.

Minerals and Railways in Central Africa

In the mineral development of Central Africa throughout the twentieth century political factors have affected the economics of mineral transport. As is well known many of the international

75

boundaries there are arbitrary, the result of a carving up of the undeveloped continent into colonial territories in advance of settlement and the full knowledge of their resources. European boundary drawing reached one of its lows of impractability with the Congo. Although eighteen times the size of England the Congo has only twenty-five miles of Atlantic seaboard and two ports, Matadi and Boma, accessible to ocean-going ships. The Rhodesias took their shape on the map by being pushed in between Portuguese territory to east and west and the Belgian and German colonies. By the first decade of the twentieth century it was already clear that, as on the Rand, the chief mineral resources of central Africa were in the main watersheds where the basement rocks of the continent were exposed. This is the case in the Bulawayo-Salisbury uplands of southern Rhodesia and still more in upper Katanga and its continuation south-eastwards into northern Rhodesia. The railway system of British Central Africa was extended from the white highlands of southern Rhodesia, across the Zambezi and up to the lead-zinc mining camp of Broken Hill. By 1909 it had reached the predecessor of the big operations to be developed on the Copper Belt twenty years later, the small copper mine at Bwana Mkubwa. It was linked across the boundary with bigger operations in Katanga. At this time the outlet for central African copper was the Mozambique port of Beira.

In the early years of the century construction was begun on lines from the Angola coast, giving a journey to Europe shorter by 3,000 miles than the Beira route. By 1911 track stretched over 200 miles inland from the port of Luanda to Malanje, but no further progress inland was made. In 1903 construction was begun on a railway from Lobito Bay, and by 1916 it already extended 300 miles from the port. Work was resumed in 1920, the line reached the Congo border in 1929 and by 1931 a link had been made through to the Copper Belt via the B.C.K. (Bas Congo-Katanga) system. The Benguela Railway, that is this line within Angola, is 838 miles in length, the Congo connecting link another 630 miles. The Benguela Railway is controlled to the extent of 90 per cent by Tanganyika Concessions ('Tanks') in which British interests are dominant. By the late 1960s £52 million had been spent on the system.

76

Another element in the transport picture was taking shape at the end of the First World War. The Belgian line which continued the Rhodesian Railway north of the border was extended to Elisabethville (now Lubumbashi) and on to Bukama, the head of navigation on the Lualaba River. In the thirties it was built on across the river

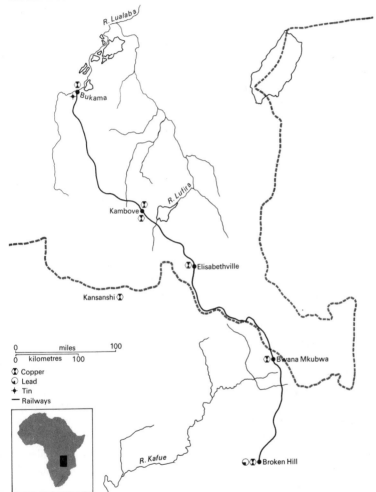

Figure 15   Mining in Katanga, the Copper Belt and at Broken Hill 1925

77

valleys of the southern Congo basin to Port Francqui on the Kasai. The Rhodesian, Benguela, and B.C.K. system now competed for traffic from Katanga and the Copper Belt. However their rivalry

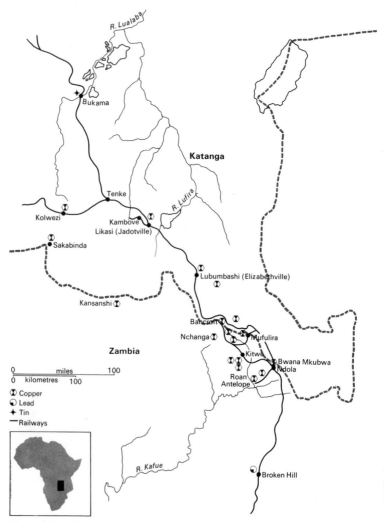

Figure 15   Mining in Katanga, the Copper Belt and at Broken Hill 1970

was by no means free. Each colonial government, and subsequently each independent state, attempted to secure traffic for its own railway system. A tripartite agreement limited the amount of Rhodesian copper which could travel to the west coast to the very small figure of 10,000 tons a year. Later still the Central African Federation wished to secure more traffic for Rhodesian Railways so that in 1960 this agreement was suspended, and after 1961 no Rhodesian copper moved to the west coast, but instead took the longer route via Beira. The unilateral declaration of independence

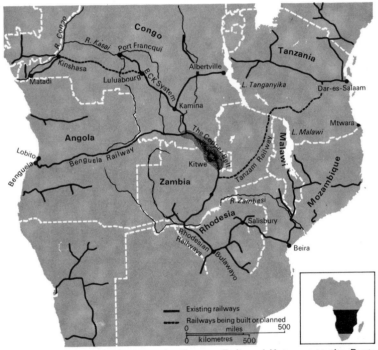

Figure 16 Railway Links from the Copper Belt and Katanga to the Ports

by Rhodesia in the autumn of 1965 transformed the situation, putting the Zambian government of President Kaunda in a very difficult position.

The Benguela Railway has again secured much of Zambia's

copper traffic and carries coal to Copper Belt smelters. Along with the Japanese, the British mining group Lonrho has proposed building a line from Port Francqui through Kikwit to Kinshasha and on to Matadi, so providing a through connection from the Congo mouth to Katanga for the first time. The cost has been estimated at £50 million in foreign exchange and almost as much from home sources. However the line would open much more of the south-western part of the Congo Basin. Meanwhile the Benguela Railway has committed itself to major expenditure to increase capacity. However Zambia was naturally reluctant to be dependent on either the Congo or Angolan outlets.

Immediately after Rhodesian U.D.I. some Zambian copper was flown out. Later road exports via the so-called Great North Road to Dar-es-Salaam became important, and it was decided to make this an all-weather road. By 1968–9 Zambian copper exports were about equally divided between three routes – the Benguela Railway to Lobito Bay, Rhodesian Railways to Beira and the road to Dar-es-Salaam. For a number of years it had been recognized that the long-term solution was a new independent railway, requiring cooperation between Zambia and Tanzania. The Tanzam railway project not only involves a costly new route but brings in another major, and perhaps eventually disturbing influence, that of communist China.

The route for the present Tanzam railway was discussed between the wars, and the practicability of a rail link from the Copper Belt to the Indian Ocean was investigated in 1952. Extension of the line which had been built inland from Mtwara into the area of the ground-nut scheme was considered. The idea of a new railway was shelved during the life of the Central African Federation and the Mtwara track was lifted. By the early 1960s a new line was being built from Dar-es-Salaam into the developing sugar producing districts of the Kilombero Valley. A long stretch of difficult country separates this from Lusaka or Kapiri on the Zambian trunk line. In 1963 the World Bank concluded that it would be uneconomic to make such a link. However, at the same time, the East African Railways made another study which disputed this conclusion. Although clearly an interested party, East African

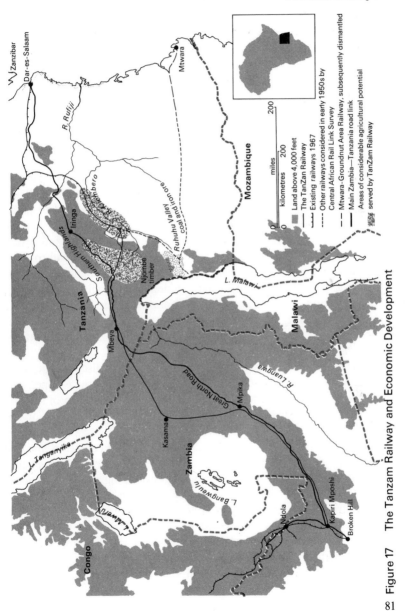

Figure 17   The Tanzam Railway and Economic Development

Land above 4,000 feet

The TanZam Railway

Existing railways 1967

Other railways considered in early 1950s by Central African Rail Link Survey

Mtwara-Groundnut Area Railway, subsequently dismantled

Main Zambia—Tanzania road link

Areas of considerable agricultural potential served by TanZam Railway

kilometres   200

miles   200

Railways pointed out that growth in Copper Belt production might make new outlets desirable. Rhodesian U.D.I. changed dreams and speculation into issues of practical concern. The capacity of the Benguela line was limited, shipment by Rhodesian Railways was considered an undesirable necessity, to be reduced as soon as possible, and, although charges to the mining companies for road shipments to Dar-es-Salaam were competitive, operating costs were high.

As early as 1964, before the U.D.I. crisis, China had expressed interest in the construction of a new African railway to the Indian Ocean. Even after U.D.I. conduct of negotiations was protracted, but by 1969, before they were completed, over 600 Chinese had finished the survey work. Under agreements signed in 1970 Chinese and local workers will together build the line, which will be financed by a Chinese loan of £169 million to be repaid by Zambia and Tanzania over a period of thirty years. Local outlays, largely for labour and materials, will amount to 52 per cent of the total cost. Construction was begun in late summer 1970 with almost 5,000 Chinese and 7,000 Tanzanian workers. At peak construction 30,000 men will be engaged in linking Dar-es-Salaam to Kapiri Mposhi. The line is to be completed by 1975. By that time the Copper Belt will have become, in President Kaunda's words, 'the rail junction of Central Africa'.[2]

5. Mineral Processing – Aluminium

After it is received from the mine an ore must be processed before it acquires a usable form. It must first be separated from its gangue. This commonly involves crushing, and one or more of various separation techniques, mechanical, chemical or electrical. Physical sorting by hand was formerly of great importance and is still practised, though supplemented and in newer operations replaced by electronic devices. Other methods include separation by gravity – the principle of the old miner's pan – by flotation, in which the valuable minerals are adhered to the bubbles of a specially chosen liquid, leaching, other chemical processes or perhaps magnetic separation.

Smelting and often also refining are further stages of production employing large furnaces, converters or electrolytic units and making heavy calls on fuel and power. Commonly the major mining concern will own smelters and refineries as well, but smaller producers may sell their concentrates to independent smelters. After this comes shaping or semi-fabricating by casting, rolling or extrusion. Ownership independent of the mine is more likely still with semi-fabrication but in some of the more fully integrated mineral industries, and outstandingly in aluminium, this stage too is controlled by the major concerns. Fabrication, the production of the finished product, is much more widely undertaken by completely independent firms spread throughout the industrial districts of the developed world. A small independent fabricator in the Black Country, in Lombardy or in the New York industrial area is the last link in the long flowline which connects a mine in central Africa or in the high Andes to the electrician, the householder, the service station of the neighbourhood of any town or city in an advanced industrial country.

With processing in all its various stages a locational choice is even more clearly involved than with mining, for nature has provided no fixed limits. The early stages of processing may take place near to the mine. This will almost certainly be the case with concentration, but in some cases there are great advantages in locating smelting operations well away from the mine. A smelter may deal with the produce of more than one mine, so that a location between a number of mines, or perhaps proximity to the market is more important. High energy consumption in smelting may be a more important locational influence than the cost of long hauls on concentrates rather than on metal. In the past availability of electricity from hydro-electric power sites has localized aluminium production in the wilder, remoter areas of advanced industrial countries. Looking at it globally, rather than at individual cases, high fuel consumption was one factor which helped to concentrate smelting operations in the coal-rich industrial countries of the north temperate zone. Even more important, and also related to their riches in coal, this is where mineral demand, entrepreneurial talent and development funds were concentrated.

In all economic activities technological change affects locational choice. It may alter the raw material supply pattern by substituting one material for another or by reducing the need for any one input. Alternatively it may change the desirable scale of the plant and therefore its locational advantages with respect to national or international markets. The location of new smelters will respond to these changing forces in so far as the company or state undertaking the development is fully aware of the situation and willing or able to act with economic rationality. If the concern has old plants it cannot respond freely to new locational influences but may expand its better plants rather than those less favoured, or, if changes in locational forces are great enough, it may close down the old and build new ones. Company structure, government policy and a host of other outside forces may affect the decision.

The inertia of plant-in-being, the persistence of old market patterns, the perpetuation of old companies and of their presuppositions and prejudices helps to preserve past patterns of smelting. Swansea which had become the world's leading copper smelter before 1800, when British mines dominated production, only gradually lost its dominance in the nineteenth century long after the leading mining districts were located outside Britain. By the early 1960s Arizona and Utah alone produced 80 per cent of U.S. copper ore but the refineries of the mid Atlantic states produced over 57 per cent of American electrolytic copper. Cornish tin output ranks Britain a poor tenth in the output of the non-communist world but it has the world's third largest tin smelting capacity. American tin consumption is more than twice that of the U.K. but the old established patterns of world smelting have proved extremely tenacious, and its smelting capacity is less than one seventieth that of Britain.

Aluminium

The principles, such as they are, the complications, the inertia but also the occasional rapid changes in the pattern of mineral processing, are all well illustrated by the world aluminium industry. One hundred years ago aluminium was a commercial nonentity of the metallurgical world. It was used with copper to form aluminium bronze and for ornaments; its other uses were also specialized. 'From the first, the jeweller and the goldsmith have seized upon it, while dentists employ it in the preparation of the pivots and the mountings of teeth, and artificial sets of teeth. It is also now used successfully in the manufacture of certain physical and optical instruments, which its lightness renders convenient and manageable.'[1] As it was one fourth the weight of silver, Napoleon III had the silver eagles of the standards of the French army replaced by eagles fashioned from aluminium. At the Paris Exposition of 1855 a bar of aluminium was exhibited – placed next to the Crown Jewels! It was impossible then to anticipate that little more than a century later aluminium would be used extensively in building and bridgework and that annual world production would far exceed that of copper and all other metals except iron. The basis of the change has been a remarkable reduction in price. Even into the 1880s aluminium was about $12 a pound but in 1969 the basic price for primary aluminium sold by Alcan outside North America was 27.5 cents a pound. Cost economy has been aided by major increase in scale of production but the basic factor was a fundamental technical change.

Aluminium metal was first made from aluminium chloride, then in the mid nineteenth century from cryolite, a double fluoride of aluminium and sodium, and by the 1870s from bauxite, the hydrated oxide of aluminium, $Al_2O_32H_2O$. It was already known that alumina, aluminium oxide, could be obtained by heating bauxite along with soda ash, but the practical bottleneck to growth of aluminium production and use lay beyond that point, in the extraction of the metal from alumina. A roundabout method employing much conventional heating was used until the eighties.

In 1886 the American Hall and the Frenchman Heroult simultaneously made the breakthrough in aluminium technology – electrolytic refining. Using a solution of alumina dissolved in molten cryolite which in turn was held in carbon lined cells or 'pots', they achieved reduction by passing a direct current through the solution. This method is still the basis of the industry today. Electrolytic refining quickly reduced the cost and widened the use of the metal. In 1890 the price fell sharply to 20 shillings or $4.8 a pound. By 1900 world production reached an estimated 7,000 tons and consumption was widening beyond the needs of jewellers, dentists and so on. By 1940 output was 0.8 million tons. In 1962 it was estimated that by 1970 world production would exceed 7 million tons.[2] In fact 1970 output was 10.0 million tons.

Depending on its richness, four to six tons of bauxite are needed to produce one ton of aluminium metal. This large loss of weight in manufacture has not localized aluminium smelting in the bauxite producing districts. Instead alumina is produced near to the bauxite workings, or alternatively at some point accessible by water transport and on the flow line between the mineral workings and the aluminium plant.

The determining factor in the location of reduction plant is the availability of low cost power, and although old plants have proved fairly tenacious, the search for power cost reduction has involved much more spectacular locational change than in most metallurgical

Table 8: Annual Aluminium production (million tons)

	1930–39	1955	1960	1969	1970
*Advanced economies**					
E.E.C.	0.12	0.32	0.48	0.85	0.91
Efta	0.06	0.19	0.31	0.77	0.82
U.S.S.R., Comecon and Yugoslavia	0.02	0.54	0.85	1.86	1.93
Other Europe	–	0.01	0.02	0.20	0.24
U.S.A.	0.09	1.40	1.80	3.44	3.61
Canada	0.03	0.55	0.68	0.98	0.96
Australia	–	–	0.01	0.13	0.18
Japan	0.01	0.06	0.13	0.57	0.73
*Underdeveloped World**					
China and Far East	–	0.02	0.09	0.17	0.17
India	–	0.01	0.02	0.13	0.16
Africa	–	–	0.04	0.16	0.16
Latin America	–	–	0.02	0.14	0.16
World Total	0.33	3.10	4.45	9.40	10.04

* Advanced and Underdeveloped defined as over or below $500 G.N.P. per capita 1966.
Sources: various.

industries. As with other metals semi-fabrication, and still more final fabrication, involves either or both an increase in weight or bulk and so has commonly been market orientated. The various parts of the aluminium industry – bauxite mining, alumina preparation, the production of primary aluminium, aluminium shaping and aluminium end use – have been frequently conducted as unintegrated operations and at widely separated locations. However, some mutual locational attraction is discernible notably in the bauxite–alumina, alumina–aluminium operations, and in its organization and structure the aluminium industry has been dominated throughout most of its history by a few, internationally-ranging concerns. These controlled the necessary patents and sometimes exercised national monopolies. Since 1945 this concentration of power has lessened, and more recently still aluminium production has spread to a much larger number of locations.

Aluminium production was long confined to a handful of advanced industrial countries. At the end of 1941 only fourteen countries had aluminium capacity and the four smallest of these had only 1.2 per cent of the world total. There was no capacity at all in the whole of Africa, Latin America, Australia or in Asia outside the Japanese Empire. By 1969 thirty-three countries produced primary aluminium and several more are shortly to join their ranks. To understand the pattern of reduction it is necessary to examine the production chain of which the aluminium plant forms the critical link.

Bauxite production

Bauxite, named after Les Baux in the lower part of the Rhône valley, remains the primary ore of aluminium though under exceptional cases, notably in the U.S.S.R., alternative mineral sources have been exploited. In the first decade of the century the world's small bauxite output came wholly from four countries – France, the U.S.A., Italy and Northern Ireland (County Antrim). In the thirties there were twenty-one producers but half the output still came from seven advanced industrial countries – the U.S.A., U.S.S.R., and Western Europe. Since the Second World War the supply situation has been revolutionized. Over the years 1934–8 France produced between one fifth and one quarter of the world supply, Hungary one eighth and the U.S.A. one ninth. The Guianas and the Dutch East Indies, then the sole important producers in the underdeveloped world, together produced rather more than France. By 1968 France alone mined more bauxite than the world had done thirty years earlier but her share of the world total had declined to only 6 per cent. U.S. output was under 4 per cent. Guyana, Surinam and Jamaica, a non-producer in the thirties, together mined 38 per cent of the world total. A massive developing trade in bauxite from the Third World to the advanced industrial countries is the dominant theme of the primary side of aluminium production. The size of this flow gives the undeveloped producers an opportunity to push the international companies to commitment to undertake further stages of production there. They have not let this opportunity slip.

Table 9: World bauxite production, 1938, 1968, 1970 (thousand metric tons)

	1938	1968	1970 (preliminary)
W. Europe	1,062	2,929	3,215
Balkans and E. Europe	1,127	6,462	7,016
U.S.S.R.	250	c. 5,000	c. 5,000
U.S.A.	316	1,692	2,052
Australia	2	4,955	9,500
China	–	c. 400	c. 400
India	15	936	1,273
Central and South America	772	19,706	24,900
Africa	–	2,873	3,300
S.E. Asia	301	1,678	2,030
World	3,849	46,661	58,686

Source: *Bureau of Mines* and *Mining Annual Review*.

In a developed economy, where skilled and semi-skilled labour and infrastructure are available, and marketing costs can be reduced, a relatively small bauxite-producing operation may still be economic. Péchiney – the leading European aluminium producer – is preparing a new bauxite operation near Mazauges in the department of Var, Provence. The deposit is only 8 million tons, the bauxite occurring in beds of $2\frac{1}{2}$ to $3\frac{1}{2}$ metres thickness beneath 150 to 300 million tons of overburden. The mineral will in fact be mined rather than quarried. This is a remarkable contrast with another project in which Péchiney has an interest – that at Boké in the hinterland of the northern section of the Guinea coast. Ultimately Boké will produce some 8 million tons of bauxite annually. Development costs are estimated at \$182 million for the mine, an 85-mile railway, a new port at Kamsar amidst the mangrove swamps, and a new township at Sangaret. The World Bank and A.I.D. (the U.S. Agency for International Development) are making loans of \$85 million for the infrastructure, the Guinea government has a 49 per cent interest in the mining company and the rest is owned by a consortium of international aluminium concerns – Alcan, Alcoa and Harvey from North America and Péchiney and Ugine-Kuhlman of France, Vereinigte Aluminium-

Werke of Germany and the Italian firm Montecatini–Edison. As with iron ore or copper in Mauritania, or indeed any base metal in the difficult environment and undeveloped areas of the Third World, mineral exploitation must be big or it cannot be justified at all.

Britain has long had interests in Ghanaian bauxite, and the experience of the Volta River power, bauxite and aluminium project – which however is separate from British Aluminium's own bauxite operations there – both confirms the economies of scale and shows that it is difficult to secure them. Great labour economies were reckoned to accrue as scale of production went up, but in spite of world growth in bauxite and aluminium production, Ghana

Table 10: Estimated bauxite production per man

Annual output of operation (tons)	Per capita output (tons)
60,000	128
400,000	635
1,000,000	1,219

Source: U.N. Studies in Economics of Industry 2.
Pre-Investment Data for the Aluminium Industry, 1966, p. 3.

has so far been unable to reap these. Her total bauxite output reached only 352,000 tons in 1966 and subsequently fell away slightly. In 1970 it was 350,000 tons.

Major new bauxite producers have been joining the lists. Thirty years ago Greece mined under 100,000 tons bauxite a year, ten years ago 750,000 tons and by 1970 2.3 million, or three quarters as much as France. Australian growth in bauxite, as in many other minerals, has been even more spectacular. Early postwar plans for the establishment of domestic aluminium production depended on imported bauxite, but subsequently huge deposits were found in the far north-east at Weipa and more recently at Gove on the northern tip of the western shore of the Gulf of Carpentaria. Ten years ago Australia had virtually no bauxite production; as late as 1966 it ranked seventh with 1.8 million tons. Output in 1968

11. *Weipa Bauxite workings, northern Queensland.*
Eight cubic yards of bauxite are excavated from the face at a time and dumped directly into the 50-ton trucks.

was 4.9 million tons, ranking it behind only Jamaica, Surinam and the U.S.S.R. In 1970 it produced 9.5 million tons, pushing ahead of all except Jamaica, and before the mid 1970s will probably be the world's leading producer of both bauxite and alumina.

The divergence between the world pattern of bauxite and aluminium production has widened, the first increasingly concentrated in the developing world, the latter still heavily localized in the advanced industrial countries. In the non-communist world only France among the industrial leaders can more than meet its own bauxite needs. There are a variety of further development possibilities: 1. improve the efficiency of large flows of unprocessed bauxite to distant reduction plants; 2. convert the bauxite into the less bulky, more transportable alumina, or 3. produce primary aluminium in the areas near to the quarries. The first two trends are pronounced, signs of the last are far fewer.

Freight charges on bauxite are quite high. As of 1965–6 bauxite from the Caribbean was priced at $7.0 to $8.0 a ton delivered to alumina plants along the U.S. Gulf Coast. Transport charges represented $3.00 to $4.50 of this total. Again assuming four tons of bauxite to one ton of aluminium – with poorer deposits the ratio

Mineral Resources

Table 11: Production of aluminium and aluminium content of domestic bauxite output in non-communist world 1968 (thousand metric tons)

	Aluminium	Aluminium equivalent of bauxite output*
U.S.A.	2,953	422
Canada	902	+
Japan	482	+
Norway	474	+
France	366	691
West Germany	257	+
Jamaica	–	2,103
Surinam	44	1,415
Australia	97	1,235
Guyana	–	930
Greece	76	440
Guinea	–	400

* assuming 4 tons bauxite = 1 ton aluminium.
+ less than 50,000 tons or negligible.

Based on: *Mining Annual Review,* 1970.

may be six to one – and putting aside additional transport charges on alumina to the reduction plant, this initial stage alone involves freight of $12 to $13 per ton of ingot. Looked at in relation to selling prices it is true that it seems a small cost factor – 0.66 cents a lb. when Alcan's basic price in the U.S. market by April 1970 was 29 cents a lb. But considered with profit margins, and taking into account the volume of production – Alcan's wholly owned plants in the U.S. in 1969 produced well over 1 million tons of aluminium – there is incentive enough for economy in movement. The internationally operating producers have therefore introduced bulk handling similar in principle, though much less spectacular than that in the iron ore trade.

Alcan has upgraded port facilities in Jamaica to permit use of bigger vessels and recently has chartered two vessels with strengthened hulls in order to extend the delivery season into the period of ice accumulation on the Saguenay River leading to its big Arvida smelter complex. In 1969 Reynolds Metals brought a 51,000 ton ore carrier into operation. British Aluminium is negotiating for the use of larger ore carriers for West Ghanaian bauxite exported from the port of Takoradi, and in September 1968 the first 50,000-ton cargo of bauxite arrived at Fos, the big new port and industrial zone near Marseilles. Although Péchiney is developing the new bauxite field at Mazauges, extension of bulk handling at Fos and its interests in Guinea bauxite and in that of Weipa points to a probable gradual French transfer from Provencal bauxite to foreign supplies, richer, more cheaply mined and more efficiently transported.

Some two to three tons of bauxite are needed to produce one ton of alumina. The industry's transport problems can therefore be reduced by locating alumina capacity in the bauxite-producing area. The advantage may be diminished by slightly higher costs for some of the factors of production, though labour will probably be cheaper. Another disadvantage from the point of view of the international aluminium company is political, the unreliability of some governments in the developing world and the opportunism of others, endeavouring to push companies with large financial interests to still more expansion. Alumina production is attractive to the Third World bauxite-producing country precisely because it makes possible a much bigger return in the way of taxes, royalties, wage payments and so on. It also lays the foundation for further advance, perhaps through to aluminium reduction.

In its Second World War expansion the American aluminium industry, whose alumina capacity had been concentrated in East St Louis, up river from the Arkansas bauxite deposits, and well placed for supplies of coal, built new alumina plants on the Gulf Coast. The next stage was to locate in the ore-producing districts. In 1966 47 per cent of Jamaica's export revenue came from bauxite and alumina, but the country was still in economic difficulties with a low per capita income ($460 in 1966 as compared with $1,620 in the U.K.) a rapidly growing population and an unemployment rate of 15 per cent. It seemed desirable to try to induce the big aluminium companies to produce more alumina in Jamaica.

Alcan pioneered Jamaican alumina production with a unit in 1952, but, from then to 1970, of the total Jamaican bauxite production of over 104 million tons, 81 million was exported unprocessed. This has earned J.$124 million in taxes and royalties.* The other 23 million tons were processed into 9 million tons of alumina, which brought in J.$72 million. That is, every ton of unprocessed bauxite earned J.$1.53 and every ton exported as alumina, J.$3.13. Additionally each ton of alumina represented much bigger local outlays for services, supplies and labour. Alumina gives yields four times as great as bauxite in foreign exchange and employment.

For several years Jamaica has only granted licences for the mining of new bauxite deposits on condition that alumina is produced. It is hoped that eventually half will be further processed in this way. In 1967 only 0.83 million tons of alumina were produced from two Alcan plants. In 1968 Alcoa began work on its first alumina plant on the island, and a year later a major alumina operation shared by three other major U.S. producers, Anaconda, Reynolds and Kaiser, was brought into production. The latter represents investment of $200 million. In 1970 Jamaican alumina production was 1.65 million tons and by the end of 1971 alumina capacity there will be 2.6 million tons.

However the location of alumina plant is not by any means tending uniformly in the direction of the bauxite-producing countries. Arvida alone produces as much alumina as Alcan's Jamaica operations. Growth in ore carrier size has reduced haulage costs so much that German aluminium producers no longer favour overseas alumina plants, and major extensions have been made recently in the Ruhr area. Some Japanese companies import alumina, but there are at least seven alumina plants in Japan and over 3 million tons of bauxite was imported in 1970.

Aluminium Reduction

Location of aluminium production is affected by the relation between scale economies and marketing and power costs. The latter is the more readily appreciated influence. At least 15,000 kWh. of electricity are required for every ton of ingot aluminium. Variations

* 1J.$ = 50p.

in power costs are therefore important and smelters have traditionally been located near to low cost power.

In mid 1970 the basic ingot price was \$649.6 a long ton. It was
estimated three years before that a difference of 1 mill (one
thousandth of a dollar) per kWh. would affect reduction costs to
the extent of at least \$15 a ton. Power costs to aluminium plants do
in fact vary by a number of mills as shown below. However, power

Table 12: Power costs to aluminium reduction plants (mills per kWh)

Canada	1.5–3.5
U.S.A.	2.0–4.0
W. Europe	4.0–6.0
Japan	2.7–8.0
Volta (Ghana)	
at 360 m.w.	4.40
at 720 m.w.	2.32
Inga (Congo)	
at 1,570 m.w.	2.50
at 25,000 m.w.	1.25

Source: U.N. Studies in Economics of Industry 2. *Pre-Investment Data for the Aluminium Industry,* 1966, pp. 15–16.

economy and new forms of generation, notably in nuclear stations,
may reduce differences within countries. In underdeveloped
countries big multi-purpose development projects providing irrigation water, flood control, improved navigation as well as power,
may make it possible to provide electricity more cheaply. The
search for cheap hydro-electric power in remote areas of the
advanced industrial countries can now be seen as merely a phase in
the development of the aluminium industry and not as a law of
nature.

Scale economies in reduction are partly related to technical
conditions and partly to the need to spread the huge overheads. A
technically optimum potline capacity is 60,000 tons. Multiples of
this size are practicable, and will bring some further production
cost reduction, but smaller units involve substantial cost increases.
In an underdeveloped country, smallness of the market may
restrict the plant to an uneconomic size. On the other hand, to
increase still more the return from its bauxite operations or alumina

12. *Aluminium pot lines, Ardal, Norway.*

capacity, an underdeveloped country may do all in its power to induce a foreign producer to build a reduction plant. Surinam brought a 60,000-ton aluminium plant into production in 1966, and Guyana and Jamaica are aiming to have their own reduction plants. However, it is said that the whole Caribbean area does not consume much more than 1,000 tons of aluminium annually. Sales into the markets of the advanced industrial countries have to reckon with their tariffs – a $7\frac{1}{2}$ per cent duty into the U.S.A. and 9 per cent in the case of E.E.C. Because of the ancilliary plant and services, essential whatever the scale of the operation, the small plant costs

Table 13: Average aluminium production costs ($ per ton, including capital charges)

Annual capacity (th. tons)	Cost of production
20	$502
30	$492
60	$463
100	$450

Source: U.N. *Pre-Investment Data for the Aluminium Industry,* 1966.

much more to build, ton for ton, than the big one. Moreover, although labour costs are lower in an underdeveloped country, more infrastructure will have to be provided, and overall development costs will therefore be greater.

Table 14: Cost of various scales of reduction plant ($ per annual ton in U.S.A.)

| Capacity | Plant type | |
(thousand tons)	Pre-baked	Soderberg
20	$1,000–1,300	$900–1,200
50	$ 750–1,050	$ 700–1,000
100	$ 650– 850	$ 650– 850
200	$ 500– 700	$ 550– 750

Source: U.N. Pre-Investment Data for the Aluminium Industry, 1966.

In the mid fifties Alcan built a 300,000 ton smelter at Kitimat, British Columbia and a 13,000 ton plant was completed at Bell Bay, Tasmania. The capital cost of the first, including power plant, settlement and port and most of the related Jamaican bauxite and alumina capacity, was £630 per annual ton. For Bell Bay, costs, excluding power plant and ore production, were £615 per ton. For the Volta River project the estimate made at the same time, covering bauxite mines, alumina plant, dam, power station and reduction plant for 210,000 tons aluminium annually, was £700. If railway, road and port expenditure is added the average reaches over £1,000 per ton.[3] Not surprisingly development has been much slower than anticipated, so that in 1970 Ghana's Tema smelter produced only 111,000 tons metal, though this represented full capacity for the unit which was eventually installed.

Aluminium production in the advanced industrial countries

The first aluminium reduction plants were coal based. In Britain, before electrolytic refining was introduced, aluminium was produced in the 1860s in two short-lived operations – one at Battersea and the other at Washington, County Durham. In the late 1870s and 1880s Oldbury was a producing centre and at the end of the eighties there were two plants on the Tyne at Wallsend and Hebburn and another at Milton in the Potteries. In 1890 the first British plant to use the Hall-Heroult process was also built to employ electricity generated from coal at Patricroft, just west of Manchester. Within four years competition from French and Swiss hydro-electric based plants had ruined these older operations, and in 1894 the British Aluminium Company was formed to build a plant based on the water power of the Falls of Foyers in the Great Glen of Scotland.

Since the 1890s there has been a continual search both internationally and nationally for lower cost power. In some countries this has led to a migration of the centre of aluminium production away from areas once attractive but where competition from less power intensive industries has forced the price of electricity beyond levels acceptable for major new aluminium plants. This trend is counter-balanced by the saving in investment from extending an old plant rather than building on a greenfield site. As a result, although some of the pioneer plants have been abandoned, others have survived as important units competing with the new locations. International power cost variations have been so great that some of the bigger concerns have chosen to build plants overseas. Finally, although electrolytic reduction is still the only significant method of production, in a considerable number of instances, especially in the postwar period, electricity has been derived from fossil fuels rather than from water.

In the immediate postwar period the search for low cost power combined with market advantages, gave North America a growing pre-eminence over the old European centres of production. In the thirties the United States and Canada averaged 36 per cent of world output, in the fifties 61 per cent. In the fifties and early sixties capacity was extended in the more favoured areas of Europe, notably Norway. In the middle and late 1960s two important new trends have become noticeable. One is a spread of aluminium production to more of the developing countries; the other, a new spurt in construction in the older industrial countries lacking major hydro-electric power resources, notably in north-western Europe and Japan. Here, alternative sources of energy – coal, natural gas

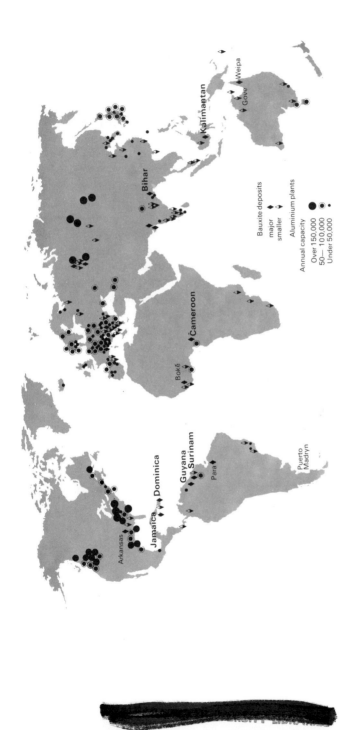

Figure 18   The World Aluminium Industry 1969-70   Minerals Yearbook, Oxford Economic Atlas, and Mining Annual Review

Bauxite deposits
major
smaller

Aluminium plants
Annual capacity
Over 150,000
50—100,000
Under 50,000

Weipa

Gove

Kalimantan

Bihar

Cameroon

Boké

Arkansas

Jamaica   Dominica

Guyana
Surinam

Para

Puerto
Madryn

and nuclear power – are being employed. Political as well as economic considerations are important. Combined with the growth of Soviet production this had already reduced the share of North America in world production to 45 per cent by 1970. North American developments are far too complex to analyse in detail here but a brief account is essential.

North American and Western European aluminium

The Hall-Heroult process was adopted in 1888 by the Pittsburgh Reduction Company (known after 1907 as Alcoa, the Aluminium Company of America). In 1891, just before the British industry migrated to Scotland, American aluminium began its peregrinations, moving out of town to New Kensington on the Allegheny River, eighteen miles above Pittsburgh. By 1894 it was building at Niagara Falls, and in 1900 and 1903 put up plants at Shawinigan Falls, where the St Maurice River falls from the Laurentian Shield, and at Massena, a power site in the remote north-west of New York State, where the drainage of the high, wild Adirondack mountains falls into the St Lawrence. In 1914 and 1915 Alcoa moved into the mid-Appalachians, an area well favoured for power generation, with plants at Alcoa near Knoxville on the Tennessee River and at Badin on the Yadkin River in north-west North Carolina.

Until the Second World War these were the only aluminium refining plants in the United States and Alcoa its only aluminium producer. However, in 1928 Alcan was hived off to take over the Alcoa interests in Canada. At this time the Arvida smelter was built on the Saguenay river well away from the industrialized belt in which Shawinigan Falls is located. Reynolds's Metals Company, formerly merely an aluminium fabricator, became the second United States primary producer in 1940, and, when the U.S. government disposed of war built plants, the Kaiser Chemical and Aluminium Corporation became the third. There were ten U.S. aluminium producers by 1970, Alcoa having about one third of total capacity.

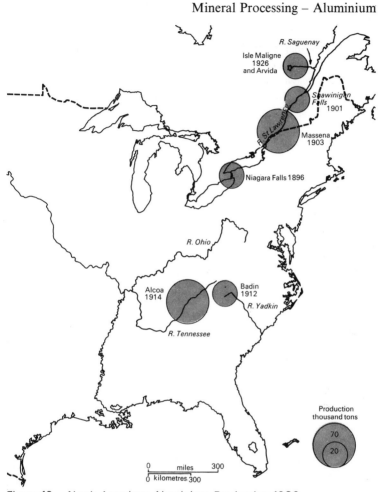

Figure 19   North American Aluminium Production 1939
Based on United States Minerals Yearbook

Expanding production, new companies, and the continuing
search for low cost power account for the gyrations of aluminium
production round the map of the U.S.A. since 1939. The industry
was then confined to four plants, all in the far east of the country,
Niagara with 11 per cent of the 1939 production, Massena with
35 per cent, Badin with 13 per cent and the Alcoa plant 41 per cent.

101

Mineral Resources

The next stage was to use the great power resources of the Columbia basin in the Pacific North-west, then being developed by the Bonneville Power Administration. The Alcoa plant at Vancouver near Portland, Oregon, was completed in 1941, the first reduction plant further west than Knoxville 2,150 miles away. The Pacific North-west now became the growth focus, and by late 1950 five plants there had 42 per cent U.S. primary aluminium capacity.

In the postwar period both eastern and north-western hydro-electric plants were extended, although the works at Niagara were abandoned in the late 1940s. However, a new element was added with reduction plants in the Gulf Coast region obtaining their power from either natural gas or lignite. By late 1961 Texas and Louisiana had four plants and one quarter of the U.S. primary aluminium capacity. A further stage of development began in 1957 when Kaiser completed a reduction plant at Ravenswood, West Virginia, supplied by a coal-fired generating plant, but with its rather high energy costs balanced by reduced marketing costs in the western part of the manufacturing belt as compared with the other much more remote units. Rival companies followed Kaiser's lead so that by 1968 three large plants were located in the Ohio basin.

Table 15: Distribution of U.S. Aluminium Reduction Plant Capacity 1939, 1950, 1961, 1968 (per cent)

	1939 (production)	1950	1961	1968	1968–70*
Appalachian hydro plants	100	43	24	22	15
Pacific North West	–	42	26	33	26
Texas/Louisiana gas and lignite	–	7	25	21	10
Ohio Basin coal	–	–	14	16	18
Others	–	9	10	9	31

* Expansion achieved or announced with expected extensions.
Based on: U.S. Bureau of Mines and Minerals Yearbook.

Major extensions continue in old areas, and new plants are still being built in the Pacific North-west in spite of its relative decline in status. The growing share of the category 'others' is connected with two new works, one at New Madrid on the Mississippi, down

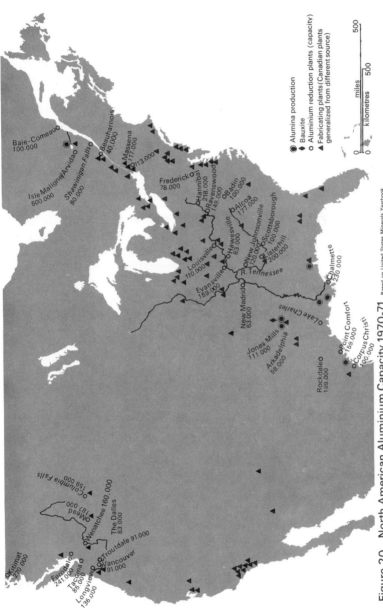

Figure 20　North American Aluminium Capacity 1970-71　Based on United States Minerals Yearbook

● Alumina production
◆ Bauxite
○ Aluminium reduction plants (capacity)
◇ Fabricating plants (Canadian plants generalized from different source)
▲ Fabricating plants

Baie-Comeau○
100.000

Isle Maligne○Arvida○
500.000

Shawinigan Falls○
80.000

Beauharnois○
40.000

Massena○
117.000
○213.000

Frederick○
78.000

Hannibal○
218.000

Ravenswood○
149.000

Badin○
100.000

Alcoa○
177.000

New Johnsonville
○120.000

Scottsborough
○100.000

Listerhill
○200.000

Louisville○
○Hawesville
170.000

Evansville○
159.000

R. Tennessee

Chalmette○
230.000

New Madrid○
63.000

Lake Charles○

Point Comfort○
159.000

Jones Mills○
111.000

Arkadelphia
58.000

Corpus Christi○
100.000

Rockdale○
199.000

Columbia Falls
○159.000

Mead○
187.000

Wenatchee○160.000

The Dalles
83.000

Troutdale○91.000

Vancouver
91.000

Kitimat○
270.000

Ferndale○
241.000

Tacoma○
85.000

Longview○
136.000

miles   500
kilometres   500
0

river from Cairo, and the other at Frederick, Maryland in the hinterland of Baltimore, at the southern end of the great markets of megalopolis.

In Canada, with the exception of the construction of the big works at Kitimat, British Columbia in the mid 1950s, locational change has been much less pronounced. On the edge of the St Lawrence lowlands, which still have the bulk of Canadian manufacturing, there are ample hydro-electric power sites. Since the mid 1950s a big new plant has been built at Baie Comeau 200 miles down the estuary from Quebec. In 1940 all Canadian production was in the St Lawrence/St John's lowland belt. In 1960 Kitimat, the only one of six plants outside this area, had 22 per cent total capacity; by 1968 28 per cent. Current expansion at Baie Comeau is again pushing up the share of the east.

In spite of its own expanding capacity the U.S.A. has become more and more dependent on imports of primary aluminium from Canada. In Western Europe and other advanced industrial countries, the need to import aluminium has been even greater. This growing need was accompanied by a changing emphasis between suppliers and finally has brought Europe to a new stage in which bigger domestic production is now emerging under rather special conditions.

Table 16: Primary aluminium. Trade Balances 1955 and 1965 (exports minus imports in thousand metric tons)

	1955	1965
Norway	+ 60.3	+232.7
France	+ 21.6	+111.0
Germany	− 39.7	−134.2
Belgium/Luxembourg	− 26.1	−114.7
United Kingdom	−262.7	−324.1
Canada	+424.9	+635.0
U.S.A.	−155.8	−293.8

Source: O.E.C.D. Non-ferrous Metals. Gaps in Technology, 1969, Table 5a.

Canada has remained a large supplier to Western Europe, though by 1965 almost half of its primary aluminium exports went to the U.S.A. France and still more Norway are the only important surplus producers within Western Europe, and although their trading surplus increased by 162,000 tons in 1955–65 this was 80,000 tons less than the increase in the trading deficit of Germany, Belgium, Luxembourg and the United Kingdom. Moreover, French production costs from hydro-electric installations are rising, and even in Norway inadequate water supply sometimes holds back production. This was the case in the works of the country's biggest producer, Ardal, in the first half of 1970. On the other hand, low-priced solid fuel supplies, natural gas or nuclear power may offer an assured regular supply at reasonably competitive prices and therefore a check on the freedom of the outsider to charge high prices. More important still, it saves foreign exchange. In Japan, coal, gas and fuel oil have been employed to support the world's most rapid build up of reduction capacity – a sixfold increase from 1958 to 1968. By 1975 Japanese reduction capacity will be 1.5 million tons as compared with less than 0.3 million in 1965.

Since 1958 France has constructed new aluminium capacity in the south-west near the Lacq gasfield, and the new Dutch industry has been located in the northern gas-producing provinces. A big smelter is planned for Sardinia in connection with what will be one of the biggest alumina plants in the world, and a 84,000-ton smelter is to be built at Amay in the Meuse valley coal district, midway between Namur and Liège. Ultimately it will probably obtain power from a nuclear station.

In 1968 new plants totalling 170,000 tons, equal to two thirds of the output of primary metal that year, were announced for West Germany. These may be extended to 500,000 tons. The scope for increased home production may be seen in the 1970 primary aluminium supply situation. Home reduction plants supplied 309,000 tons ingot; 400,000 tons were imported. Some units will be in well-placed coastal locations, as at Hamburg and Stade, a little to the west, a site which qualifies for the graded Saxon state assistance for industrial development. Even before this round of expansion Germany had aluminium production on the Rhine, in the Ruhr coal and lignite area. Two of the new plants will be in the Ruhr, helping along the impressive diversification of that area's traditional heavy trade, one at Dinslaken on the Lippe Canal to the north-west, the other at Essen. The Essen plant shows the range of influences which may complicate a location choice, rendering it very different from the traditional narrow 'economic' decision. The 84,000-ton plant of Leichtmetallgesellschaft – which may eventually be extended to the extraordinary size of 260,000 to 500,000 tons – will occupy a 370-acre site made cheaply available by the city of Essen, which is eager to find new employment. It will benefit from a guaranteed investment subsidy under local coal-consumption laws, and will get extra tax concessions on 15 per cent of the total investment costs. The case of British aluminium development is even more a reversal of past conditions. It is considered more fully in Chapter 9.

Aluminium production in the developing world

Outside the advanced industrial countries, processing has made much less progress. Recently the Australian Mining Industry Council has pointed out that there are serious problems even in such an affluent society, where labour is skilled and technical expertise is available. Australia has high transport and power costs. There are strongly competing demands for capital, with returns in mineral development greater than in mineral processing. It has been reckoned that if her planned growth in bauxite production to 1975 was processed at home the power needed would be the equivalent of four Snowy River power schemes. Finally, Australia lacks adequate markets – so much so that in the case of a metal which competes strongly with aluminium it is said that, even after allowing for freight costs, Japanese copper smelters can offer higher prices for Australian concentrates than home producers.[4]

In other poor countries, rather than rapidly developing ones like Australia, the balance of considerations differs in various respects. Capital and transport costs will still be high but labour is scarce and untrained and therefore likely to be expensive in real terms. On the other hand, they may have potentially cheap power and certainly, with rapid decolonization and the assertion of the rights of formerly subservient nations, may have ambitions for both the revenue and the prestige of industrial development. The Volta River Scheme in Ghana proves that the path to a fully-fledged aluminium operation may be far from smooth.

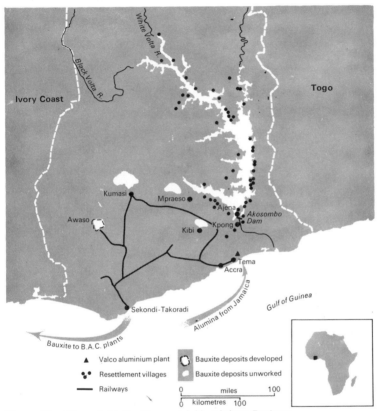

Figure 21   Ghana's Volta River and Aluminium Projects

Hydro-electric power generation on the Volta River was first suggested in 1915, and by 1924 the Gold Coast government was entertaining the first ideas about aluminium reduction. More serious analysis of prospects was undertaken by a South African engineer in 1938 with suggestions of a dam, power generation and reduction plant at Ajena, using bauxite from west of Kumasi and

107

from Mpraeso, midway between Kumasi and the dam. In 1942 British Aluminium began to mine bauxite at Awaso south of Kumasi for use in Britain but by 1951, along with Alcan, it had concluded that reduction of aluminium in Ghana was practicable. A year later the British government was involved. In the years 1949–54 British aluminium fabricating capacity had increased by one quarter and aluminium imports were rising, four fifths of British total supply coming from dollar sources. In 1951 95 per cent of imports were from Canada. The British government therefore expressed keen interest in the Volta scheme. Overall costs of bauxite operations, dam, reduction plant and the accompanying public works were estimated at £100 million for a production of 80,000 tons of aluminium, or £144 million when the ultimate capacity of 210,000 tons was attained – an impressive example of the investment economies of scale.[5] Before the project received official approval a Preparatory Commission spent 1953–5 in a very full inquiry into all aspects of it, including the effects of the great reservoir on the area, on other aspects of the economy and on current and prospective social and health conditions. In 1956 Alcan declined the offer of a long-term power contract. Ghana became independent in the following year and President Nkrumah decided to push on with the scheme. The United States was asked to help, and agreed in principle, in spite of current overcapacity at home, for it did not wish to alienate the new Africa or to repeat a situation like that at Aswan where the Russians had stepped in when Western support was withdrawn. Ghana was put into contact with the Kaiser Company which reassessed the engineering situation, and recommended three important cost savings – the dam should be at Akosombo not Ajena, the newly built port of Tema should be preferred over the dam as the smelter location and, in the early stages, bauxite should be imported. At the end of 1959 the Volta Aluminium Company (V.A.L.C.O.) was organized with investment by Kaiser Aluminium and a minority interest by Reynolds Metals. Work already had begun on the Akosombo dam, and it was inaugurated in 1965. The first V.A.L.C.O. aluminium was sold in the spring of 1967. The whole project is of wide significance. The dam, power station and transmission system cost some

£56 million, Tema as a new port and industrial centre £35 million, and V.A.L.C.O.'s investments, some £60 million. There have been implications for fishing, navigation, irrigation, the resettlement of 80,000 people from 600 towns and villages in the area of the reservoir in fifty-two new villages, and the provision of power at home and for sale to neighbouring countries for innumerable other industrial and commercial purposes.[6] The Ghanaian government still hopes that bauxite will be mined locally for V.A.L.C.O. with the establishment of a big alumina industry. Kaiser has made an analysis of this which shows that capital investment would be enormous and is apparently not available. Doubts about the country's political stability and attitude to nationalization of mining is another retarding factor. Meanwhile V.A.L.C.O. imports all its alumina from Jamaica while British Aluminium continues to export western district bauxite through Takoradi.

Ghana's course to major aluminium production has been long, chequered and not altogether successful. After huge investment the industry is still unbalanced. On the other hand, V.A.L.C.O. does benefit from economies of scale and by 1972 capacity is to be expanded from the present 110,000 tons to 145,000 tons. In other developing countries the position is on many counts less favourable.

By 1970 India had six smelters with a combined capacity of no more than 168,500 metric tons. Of this 80,000 tons was in the plant at Renukut near Rihand in Uttar Pradesh leaving the other five with an average capacity less than 18,000 tons. Four years ago India's plants ranged in size from as little as 2,500 to 10,000 tons, and it was then said that in a 10,000-ton plant production costs were 20 per cent more than the cost, insurance, freight, that is, the delivered price, for imported aluminium. Expansion to 20,000 tons would be necessary to meet import prices without protection.[7] However, perhaps protection, high cost and low output production may be justified in this case for the growth prospects are so great. As late as 1961, when U.K. per capita consumption of aluminium was 5.4. kgs (U.S. 9.8 kgs), the figure for India was as low as 0.07 kgs. Since then India's consumption has grown very rapidly because of large power development, the displacement of copper in the electrical industries, and of other non-ferrous metals. In 1968 her rated aluminium capacity was 118,500 tons; by 1973–4 it is estimated as likely to be 345,000 tons. Most of the increase will be at new plants but the two smallest plants of 1968 will also have substantially increased in size. Indian planners deal with a nation of 540 million which, in spite of the many set backs of the 1960s, offers tremendous growth prospects. Expansion of aluminium reduction elsewhere cannot always be so easily justified by potential demand if not by current economics. This is highlighted by the situation in Latin America.

In 1966 Surinam, Brazil and Mexico were the only aluminium producers in Latin America with a total production of 81,000 tons. By 1969 Venezuela had joined them and output had risen to 143,000 tons. Most of Latin America's plants are small and have high costs.

Late in 1967 the Caroni plant in the Guyana region of Venezuela came into production with a capacity of 10,000 tons. It is now to be expanded to 22,500. Brazil completed a 25,000-ton smelter at Pocos de Caldas; Alcan has a 27,000-ton smelter at Ouro Preto and is building a 10,000 ton a year unit near Salvador (with a medium term expansion target of 50,000 tons). For the two completed units there are advantages to compensate for their smallness, notably nearness to the Belo Horizonte–São Paulo industrial area, local bauxite and fairly cheap power. In 1969 Argentina decided on an aluminium project following completely different principles. A reduction plant is to be built at Puerto Madryn on the Patagonian coast, north of the River Chubut, 650 miles south-west of Buenos Aires. Whereas the other Latin American countries have built small units, Argentina is going for scale economies. Even in its expanded form Venezuela's Caroni plant has cost some $43 million for a capacity of 22,500 tons; Puerto Madryn will cost $120 million through to 1974/5 but its capacity will be 154,000 to 165,000 tons. At the moment Puerto Madryn is a town of about 5,000 people, mostly of Welsh origin. At small cost its port will be able to handle alumina carriers of 50,000 tons. Power will be brought from installations costing $63 million on the River Fataleufu, 500 km. away in the Andes. According to Argentinian sources, a new low cost for Latin America aluminium production will be established, the metal being delivered in Buenos Aires for as little as $495 to $500 a ton as opposed to $600 to $678 which – according to the Argentinians – is the production cost in Mexico, $600 to $648 in Venezuela and an extraordinary $811 to $850 in Brazil. Foreign exchange will be saved on Argentina's considerable and rapidly growing aluminium imports (40,000 tons in 1968, an estimated 105,000 to 120,000 in 1980) and, at the anticipated cost, a profitable export business may be built up with other members of the Latin American Free Trade Association. However, the member countries are almost all poor, their markets are scattered and many of them have their own expansion aspirations. Experience already suggests that it would be wrong to expect too much of L.A.F.T.A. and moreover the reported Puerto Madryn figures seem rather too rosy to be wholly convincing.[8]

Aluminium production is being promoted elsewhere in the developing world, largely with the intention of supplying markets in the advanced economies. Greece, the Philippines, Puerto Rico and Bahrain – the last employing natural gas which until now has largely been flared away – are important cases. Some of them hope to build up home fabricating as well. No doubt there will be some success but development plans suggest that the advanced industrial countries will retain and consolidate their overwhelming predominance in aluminium reduction.

6. Copper

Copper had always been the world's leading non-ferrous metal until surpassed by aluminium in the late 1950s. By 1969 output was 5.9 million metric tons as compared with 9.4 million for aluminium. Copper has been plagued with substitution by other materials, wide price variations and a great deal of political intervention. Even so, by 1975 mine output in the non-communist world is expected to be 2 million tons or almost 40 per cent more than in 1970.

Copper and bronze, its alloy with tin, were used in Egypt by 3,500 B.C., in Mesopotamia by 3,000 B.C. and in China at least as early as 1,500 B.C. From that time on a strange romance has always attached to copper mining, a romance which seems to have been absent with, for instance, zinc or lead. It is true that the summer holiday-maker, dismayed by the Wylfa nuclear power station or the new Holyhead aluminium smelter, may not readily recognize the romance of the little town of Amlwch, but it is there, if he probes a little, and it came about as the result of copper mining. In 1766 there were six houses, inhabited by fishermen. On 2 March 1768, a date subsequently celebrated in the neighbourhood by an annual festival, rich copper ores were discovered within seven feet of the surface of the quartz and shale mass of Parys mountain two miles to the south. The Parys and Mona quarries became true copper bonanzas with an output worth £300,000 a year at their peak. By 1801 the huddle of fishermen's cottages had become a town of 6,000, with 1,500 working in the mines and twenty vessels carrying ore to Lancashire smelters. Output fell off sharply after the first decade of the nineteenth century but continued in smaller volume for more than half a century. A century after the ore discovery, though described as 'dingy and disagreeable', the town had over 800 houses. Falling prices in the 1870s brought the end of

Table 17: World Copper: Production and Consumption (million tons)

	1955	1960	1965	1969
Mine Production				
Congo	0.23	0.30	0.28	0.36
Canada	0.29	0.40	0.45	0.50
Chile	0.43	0.52	0.59	0.67
Zambia	0.35	0.57	0.68	0.70
Australia	0.04	0.10	0.09	0.12
Japan	0.07	0.09	0.10	0.12
Mexico	0.05	0.06	0.06	0.06
Peru	0.04	0.18	0.18	0.21
Philippines	0.02	0.04	0.06	0.11
South and South-west Africa	0.06	0.07	0.10	0.15
Others	0.20	0.25	0.29	0.35
Sino-Soviet Bloc	0.37	0.54	0.88	1.04
U.S.	0.90	0.97	1.21	1.40
World	3.07	4.09	4.98	5.75
Consumption				
Europe	1.13	1.58	1.72	1.81
Africa	0.02	0.03	0.04	0.04
America (other than U.S.A.)	0.21	0.20	0.35	0.41
Asia	0.10	0.31	0.44	0.76
Australasia	0.05	0.07	0.08	0.09
Sino-Soviet Bloc	0.37	0.65	0.90	1.10
U.S.A.	1.16	1.00	1.50	1.54
World	3.06	3.85	5.04	5.75
L.M.E. average prices: £ a ton	n.a.	£245.8	£468.8	£620.8

Source: British Metal Corporation *Annual Review of Non-ferrous Metals.*

mining, and by 1891 Amlwch's population was less than 4,500. With innumerable variations this theme of boom and decline was repeated time after time in copper mining districts the world over. Men, entrepreneurs and capital flowed from old to new mineral districts.

Into the nineteenth century copper and bronze still retained their time-honoured uses – for artistic products, hardware, cannon and for the sheathing of vessels. Annual average world production in the first decade of the nineteenth century has been estimated at under 10,000 tons or little more than half a day's present output. Even as late as the 1850s only 50,000 tons were produced each year. After that the growth of the electrical industry – the telegraph, lighting, electricity generating and transmission – boosted consumption. Annual production in the nineties reached 370,000 tons. By the mid 1930s the average was 1.7 million.

With growth, the centres of production and the types of ore mined have both changed. Half the world's output in 1830 came from British mines, particularly those of Cornwall. By 1900 her output was less than 0.2 per cent of the world total. Chile was the world leader in the early 1860s, Spain occupied the first place briefly, but by the mid 1880s the United States was dominant and has retained its lead ever since. But, whereas in 1900 its share was over 57 per cent of the world total, by the late thirties it was under half and by the late sixties less than one quarter. It seems probable that in the seventies the U.S.S.R. will move into first place. Over the last two hundred years there has been a shift from rich to leaner ores, accompanied in the last seventy-five years by a switch from shaft to open-pit mining and by very great increases in the scale of the average operation. In the eighteenth century some British copper ores contained 13 per cent metal. The average content of the ores raised by the Calumet and Hecla Mining Company in the Keweenaw Peninsula of Upper Michigan, U.S.A., in the 1880s was an astounding 20 per cent, and sometimes great masses of almost pure copper were encountered. At the start of the twentieth century most smelters would still not work with ores of lower than 10 per cent copper. By 1940 the technical revolution in the industry was well advanced and 1 per cent ores were reckoned reasonable: in the mid fifties less than 0.7 per cent was acceptable. On the other hand it may be worth emphasizing that whereas a 0.7 per cent ore may be profitable in a well-developed, stable economy a higher ore content or a very big ore body would be a prerequisite for development in an underdeveloped one. At the end of 1969 an agreement was signed for the opening of the 500 million ton Cuajone copper deposit in Peru. Development costs are estimated at \$355 million. The ore grades slightly over 1 per cent copper.

Most of the early copper mines were shaft operations working vein deposits. Their reserves were usually small – of one million tons or less. Because of the high cost of shaft working only rich deposits could be profitable. Between the pockets of rich ore were areas of barren ground, so that unless care was taken to leave a balancing reserve of rich ore – the 'picking of the eye of the mine' referred to earlier – profit made in good times could easily be dissipated in driving through unprofitable sections looking for another vein. Elsewhere solutions ascending from a cooling magma penetrated porous rocks such as sandstones, impregnating them with copper – the so-called hydrothermal method of ore formation. The result was a porphyry ore deposit or, more descriptively, a disseminated ore body. As will be understood from this method of formation these deposits are big but low in grade. The chief examples are in the western states of the U.S.A., in the Andes and the Kounrad deposit in the desert region of Kazakhstan. A third method of formation involves the laying down of copper rich sediments over large areas to form distinct beds – so-called stratiform deposits. Cases of this are found in the copper producing areas of central Africa and Dzhezkazgan, also in Kazakhstan. In mining characteristics, though not in formation, these deposits are not materially different from the porphyry coppers. In both mass mining is applied, whether by underground methods or by open pit. Caving or stripping, shovelling, transport and concentration all have to be on a big scale to make the operation profitable. The results have been spectacular. They were summed up by the American mining engineer, J. H. Hammond.

So long as mining was based upon rich vein deposits ore reserves were small; it could hardly be otherwise. Entirely aside from the difficulties involved in finding and opening large mines where the bodies themselves were small, it was not economically advisable to pay out in advance the amounts necessary for underground work to develop large ore reserves, and it would rarely have been possible technically to hold the ground open without unwarranted expense until it came to be worked. Copper mines with few exceptions were . . . hand to mouth affairs. They had to pay from the grass roots, for no one could give the sound assurance necessary to warrant investment of large sums of capital. In the main they were speculations, not business enterprises.[1]

Along with the development of the very different technique of mass mining went the introduction of new flotation techniques, improved smelting and refining. Together these advances revolutionized world copper production. The change first became obvious in the United States.

U.S. Copper production

The first Michigan copper mines were opened in 1844–5. They soon took the leadership from smaller eastern mines. The Michigan mines were all located in a remarkable belt of ore, 100 miles long by four miles wide, running the length of the Keweenaw peninsula which juts out into Lake Superior. Even before the porphyry mines were opened Michigan was losing out to bigger producers further west. The Keweenaw shafts were deep and, in spite of the occasional blocks of native copper, margins narrowed as the journey to the face lengthened. In 1871–5 Michigan produced 89 per cent of the United States output; by 1887 Montana was ahead and by 1901–5 Montana and Arizona together produced 60 per cent to Michigan's 26 per cent. By the late 1890s the first moves were being taken to open the porphyries, and new pits in Arizona, Nevada, New Mexico and Utah were soon developed. Arizona was the leading copper state by 1910. It retains that lead with more than half the United States output of mine copper in the late 1960s. However the largest single producer is in Utah, the Bingham Canyon pit in the desert west of Salt Lake City.

At the beginning of the century Bingham Canyon was a narrow, scrub-covered valley leading up to a hummocky mountain from which gold, silver and lead had already been mined. The valley was lined with dingy, impermanent looking, frontier-type houses and shacks. By 1900 it was realized that the mountain block contained massive disseminated copper deposits which, if worked open-pit.

116

13. *Bingham Canyon, 1909.*
The early levels of the Utah Copper Company's operations.

14. *Bingham Canyon today.*
The hill mass of 1909 has become what has been claimed as the biggest man-made excavation on the earth's surface. The scale of the pit can be appreciated by noticing the trains on numerous levels.

and on a very big scale to supply a local smelter, could be mined successfully. Yet the riskiness of the venture may be gauged by the fact that it was said that at that time in Butte, Montana, better material was being thrown away as waste. Production began in 1907, and the mountain has since been replaced by a pit claimed as the world's biggest man-made excavation, 2,500 million tons overburden and ore having been removed. Bingham Canyon ore now grades an average of 0.75 per cent copper.[2]

As is frequently the case with mining, though old districts decline they do not fade out completely. Among the ghost towns and abandoned pits of the Keweenaw peninsula production still goes on. Important new discoveries are occasionally made. In 1966 a new 35 million ton, 1.5 per cent copper ore body was proved, and 1968 Michigan production was 68,000 tons of copper, 6.2. per cent of the United States total. Even so the Keweenaw peninsula is a problem area.

The opening of the western porphyries pushed the U.S.A. to the peak of its pre-eminence in world copper, but American and

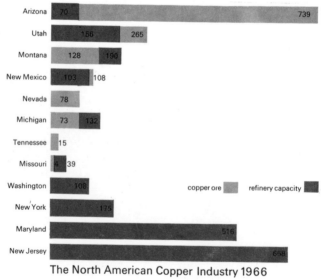

The North American Copper Industry 1966

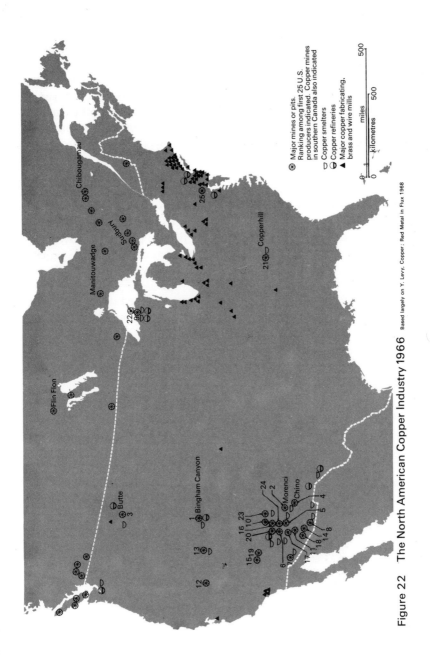

Figure 22   The North American Copper Industry 1966   Based largely on Y. Levy, Copper: Red Metal in Flux 1968

Major mines or pits.
Ranking among first 25 U.S.
producers indicated. Copper mines
in southern Canada also indicated
Copper smelters
Copper refineries
Major copper fabricating,
brass and wire mills

miles   500
kilometres   500

Chibougamau
Manitouwadge
Sudbury
Copperhill
Flin Flon
Butte
Bingham Canyon
Morenci
Chino

Mineral Resources

European companies soon invested heavily in similar deposits in Chile and Peru, in the Belgian Congo and in Northern Rhodesia. In the 1934–8 period Soviet production, though increased by all the resources of the state, averaged well under 100,000 tons or about one sixth of U.S. levels. Twenty years later at 450,000 tons it was a little under half the U.S. total and the U.S.S.R. had risen from sixth to fourth rank among world producers. By 1970 with an output 58 per cent that of the United States it ranked second in the world. In 1967 Chile, Peru, Zambia and the Congo, the leading copper countries of the Third World, produced 1.97 million tons or 39 per cent more than the U.S.A. In 1906–7, when only the first two were producers, they had mined only 40,000 tons or one tenth the U.S.A. output. Looking at it in another way, since then U.S. output has increased two and a half times but the output of the four is now just short of fifty times the 1906–7 level. Copper mining there developed under difficult conditions of access and environment and under either a directly colonial or an economically colonial regime. Over the last ten years their situation has changed radically.

Copper and the Latin American economies

In the mid sixties Latin America possessed 30 per cent of the world's known reserves of copper: in 1970 its only three significant producers, Chile, Peru and Mexico together turned out 15 per cent of the world's total. In lead and zinc too Latin America's production was less than its share of world resources, but in tin rather more. In general its mining development has lagged behind that of the world as a whole over the last fifteen years, though there have been notable exceptions, for instance, Peru. There are a number of reasons for this relatively slow development. Latin America lacks indigenous capital or entrepreneurial ability. American, European and Japanese companies have invested elsewhere, where political stability is reckoned greater and where returns on investment seem more assured.

In few Latin American economies does mining provide a significant proportion of the G.D.P. On the other hand it has brought in

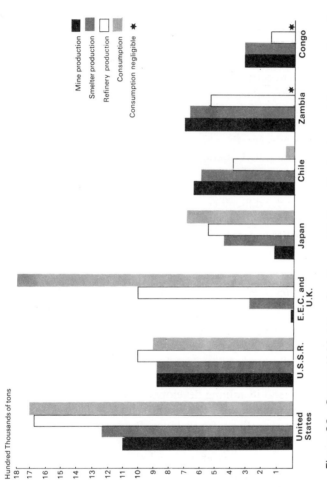

Hundred Thousands of tons

Mine production
Smelter production
Refinery production
Consumption
Consumption negligible ✱

United States U.S.S.R. E.E.C. and U.K. Japan Chile Zambia Congo

Figure 23   Copper 1968, Mine, Smelter and Refinery Production and Consumption of leading countries   Based on Metal Bulletin Handbook

Mineral Resources

Table 18: Role of mining in G.D.P. and employment in Latin America (*estimates*)

	Share of G.D.P.* 1958	1967	% of labour force in mines† 1965
Argentina	0.4	0.4	0.3
Bolivia	8.0	9.0	3.9
Brazil	0.2	0.5	0.4
Chile	7.0	6.5	3.5
Colombia	3.7	3.4	1.1
Mexico	2.3	1.5	0.8
Nicaragua	1.8	1.8	?
Peru	4.0	4.2	1.6
Venezuela	1.8	1.3	0.7

* Except for Colombia excludes petroleum.
† Excludes petroleum.

Source: 'Mining in Latin America', *Economic Bulletin for Latin America*, XIV, 2, 1969.

Table 19: Percentage of world copper, lead, zinc and tin production from the leading Latin American producers 1934–8, 1956–60, 1969, 1970

	1934–8	1956–60	1969	1970
Copper				
World output (m tons)	*1.7*	*3.6*	*5.8*	*6.3*
Chile	20%	13.6%	11.6%	10.6%
Mexico	2.3%	1.6%	1.1%	1.1%
Peru	2.0%	2.8%	3.5%	3.3%
Lead				
World output (m tons)	*1.5*	*2.3*	*3.1* (1968)	*3.4*
Mexico	13.5%	8.7%	5.2%	5.1%
Peru	2.6%	5.4%	5.3%	4.8%
Zinc				
World output (m tons)	*1.6*	*3.2*	*5.1* (1968)	*5.4*
Mexico	9.0%	7.8%	4.7%	4.6%
Peru	0.6%	4.6%	5.9%	6.1%
Tin				
World output (m tons)	*0.16*	*0.17*	*0.20*	*0.21*
Bolivia	15.6%	14.1%	14.5%	14.2%

Based on: W. R. Jones and *Mining Annual Review*, 1970.

foreign capital, caused developments in infrastructure and in manufacturing which have wide multiplier effects, and made a very much more important contribution to export revenue – 20 per cent for Latin America as a whole but over half in Peru and more than three quarters in Chile and Bolivia. In Peru and Chile copper is the chief mineral product. American interests have dominated their mineral production, an estimated 46 per cent of U.S. investment of $1,100 million in Latin American minerals being in Chile and 24 per cent in Peru.[3] Recently the governments of both countries have taken steps to control its own copper industries.

'There seems to be no gold here' is said to have been the comment of a disillusioned Diego de Almagro, the companion of Pizarro, who in 1535–8 was the first European prospector of the riches of Chile. The country's great wealth lay elsewhere. From shortly after the Spanish Conquest small tonnages of copper were produced, and in the years 1860–64 Chilean production ranged from 60 to 67 per cent of the world total. There was some investment of British capital in the 1870s but by this time the country's importance was declining, and 1899 output was less than half that of thirty years before. By this time nitrate was far and away the most important Chilean mineral product. As atmospheric fixation and then bulk production methods for ammonia were introduced in the advanced industrial countries, nitrate production in turn dwindled and copper came into its own again.

Soon after the U.S. western porphyries were opened up it was recognized that Chile had a number of similar deposits. Big developments were undertaken in the dozen years following 1912. In 1907 only 27,000 tons of copper or 3.7 per cent of the world total came from Chile but by 1924 the output was 190,000 tons or 13.8 per cent of the world figure. Chilean copper output in 1969 was 11.6 per cent of a much expanded world total, providing 80 per cent of Chilean export revenue. There are many mines, but three locations have dominated twentieth-century development: Chuquicamata in the north, Potrerillos–El Salvador, 200 miles south-east of Antofagasta, and El Teniente 530 miles farther south.

The first of the big new twentieth-century operations was the Braden field, better known as El Teniente, south-east of Santiago.

15. *Chuquicamata Copper Mine.*
Note the extensiveness of mining and processing and the workers' settlement
starkly outlined in a barren environment.

Here operations take place within over 2,000 feet of a vitrified old volcanic plug in what must surely be one of the most romantic and yet most difficult sites for mining in the world. The mine town and some of the essential plant is strung out across the extremely steep outer slope of the old volcanic mass, and there is so little flat land that mine waste must be carried away for dumping. Access is difficult. In the nineteenth century small quantities of ore from El Teniente were sent to Swansea smelters after a packhorse journey to the coast. Before large-scale mining could begin in 1912, machinery had to be hauled in by teams made up of a total of 2,500 oxen.

A year before Braden was brought into production an even bigger deposit was surveyed on the Atacama desert edge of the western cordillera. Fortunately the Antofagasta and Bolivian Railway already ran within five miles of the ore body. By 1916 the

Chuquicamata open pit was in production. In the mid 1960s it employed 7,000 workers. The third big producer was at Potrerillos and here, as the better ore was worked out, a near-by ore body, appropriately named El Salvador, was opened to make use of the Potrerillos facilities. Together in 1967 these three major mines produced 550,000 metric tons or 79 per cent of the Chilean total. All these were North American owned. Anaconda, with its origins in the mine at Butte, Montana controlled Chuquicamata and Potrerillos–El Salvador, Kennecott Copper and the El Teniente mine. For long they acted in the common way for expatriate companies, a way matched by the British and Belgian concerns of Central Africa. This involved remittance of much of their profit to shareholders at home while doing relatively little – to the 'host' country apparently very little – to build up the prosperity of the mineral producer.

16. *El Teniente Copper Mine.*
The community of Sewell on the extremely steep slopes within which El Teniente copper is mined.

Mineral Resources

In the early postwar years Chile's status as a copper producer declined. Output in the world went up 52 per cent between 1947 and 1958, in Northern Rhodesia by 103 per cent but Chilean production rose by no more than 8 per cent. At the same time the proportion of her output refined at home fell. Until the mid fifties there were very good prospects for copper, but the U.S. firms were not willing to make the major investment to expand output. It was said that costs were too great and that the Chilean business climate was unsuitable for investment – wages and social benefits were high and there was too much propensity to strike. At the end of the fifties demand for base metals fell sharply away, and there was agitation in the western U.S.A. for more restrictions on mineral imports which would clearly adversely affect the income of Latin America. The Chilean attitude was summed up in the remarks of R. Tomic in the Chilean Senate in 1961. He described the operations of the two big U.S. groups as '... mere conduits of two enormous commercial enterprises of world-wide importance, that produce huge quantities of copper in the U.S. and other centres and have, or can at any moment have, similar and even competing interests in other parts of the world. Their mining, industrial and commercial interests may well be antagonistic to those of Chile.'[4]

In the course of the sixties it has been recognized that Chilean copper reserves must be the chief support for the much needed economic growth and social revolution. At the same time Chile has taken the initiative in gaining an increasing and eventually a predominating control over the big international concerns. The picture which Tomic painted less than a decade ago is now outmoded.

Although progressive by Latin American standards, Chile is still a poor country. In 1960 the wages of more than half of all its workers were reckoned insufficient to provide a minimum acceptable standard of living, though definition of this standard is notoriously difficult. Big estates, inefficiently operated, provided an inadequate food supply. In the mid sixties population was increasing by 3 per cent a year, food output by only 2 per cent and already $85 million was spent annually on imports of basic foodstuffs. Higher average income levels and land reform were vital but in
126

turn they required large capital investment and encouraged inflation.

In the three years from 1964 Chilean wages went up generally by over 40 per cent and in farming by 70 per cent. As a result the pace of inflation was frightful – a rise in prices of 45 per cent in 1963, 38 per cent in 1964 and a little over 25 per cent in 1965. At this point President Frei tried to push through a long delayed land and agricultural reform programme. His Land Reform Bill provided for the expropriation of almost all estates of over 200 acres, in order to create 100,000 new small-holdings and increase agricultural output by 36 per cent by 1970. Partly because of drought, partly because of growing opposition from the right, the programme made only slow progress. Copper revenue was designed to finance this reform. To this end production was to be increased and ownership reorganized.

The copper law provided for 'Chileanization' of mining, that is a government share in the American concerns, and for a doubling of output to 1970. In some of the operations the Chilean government became a majority shareholder and in other cases was content with a minority share. El Teniente is an example of the first, but Anaconda was allowed to retain 75 per cent of the interest in the new Exotica ore body near to Chuquicamata. Similar retention of American control was negotiated for the Rio Blanco project of the Cerro Corporation. Chilean copper revenue was helped by a series of interruptions to world copper supplies – the effect of Rhodesian U.D.I. on Zambian production, and then the nine months U.S. copper strike in 1967–8 – but the copper law ran into difficulties.

Major new extensions have been made at El Teniente and the 100,000 ton Exotica mine was planned for production in 1970. Even so 1970 output was a little under 700,000 tons, no more than 6 per cent greater than the level of 1966. More attention is to be given to medium and small mines in order to spread the effects of employment and high mining wages. From 1968–72 capacity at the big mines will increase from 570,000 to a little over one million tons a year. In the same period, medium and small mine capacity is to go up from 150,000 to 300,000 tons. The competitiveness of this group will be increased by the establishment of a chain of regional ore treatment plants designed to create a small but valuable number of new jobs in areas with high unemployment such as Coquimbo, Antofagasta and Atacama. In the course of 1970 further plans for regional reduction were announced. Overall there has been a steady increase in the amount of copper refined in Chile – a process encouraged by a change in the tax law. Twenty-eight per cent of output was refined at home in 1964, but 78 per cent by 1969.

In other respects things have gone far from smoothly. Strikes in the mines have cut output and forced up costs. The Senate tried to introduce provision for lavish benefits for mineworkers into the copper law. When these were rejected there was a five-week strike at the end of 1965 which cost 61,000 tons production, £6 million in taxes from the American companies and £5 million in export revenue. This was followed by a three-month strike at two mines which ended with the granting of a 25 per cent wage increase and loans and bonuses of some $225 per worker – including a bonus for 'strike settlement'. There were major fringe benefits – extras for night work, school allowances for children and improvements to housing. Other strikes have followed – a twenty-one day stoppage in early autumn 1970 at Chuquicamata cost Chile $750,000 a day in exports. When this was settled the unions secured a 47 per cent wage increase and a 'cost of strike' bonus. In these circumstances it is not surprising that inflation has continued; in 1969–70 the cost of living rose by 30 per cent. Chile's economic situation is no easier. In 1969, when about $200 million worth of food was imported, copper exports were valued at over $960 million or 75 per cent of the total. In short, copper is vital in the short term as the source of exchange for purchases of food and other foreign goods, and in the long term to finance the overhauling of the economy to make further food imports unnecessary. In the course of 1970 and 1971 there were further steps in control over copper mining.

At the beginning of the year the Anaconda mines of Chuqui- camata and El Salvador were in effect nationalized. Compensation for 51 per cent of the stock which the state acquired was to be paid in semi-annual payments over twelve years, payment for the rest being related to copper prices. Anaconda was to receive 1 per cent of gross sales and a management fee. It retained its 75 per cent

128

interest in Exotica. In September 1970 the Marxist, Dr Salvador Allende, was elected President. President Allende was committed to full nationalization in place of the 'Chileanization' of Frei and by summer 1971 had carried this through. If compensation is paid in Chilean government bonds, it is recognized that rapid inflation will make their compensation valueless in a few years. In the case of Anaconda as much as 60 per cent of its income now comes from Chile. Complete expropriation of U.S. copper interests removes $300 million of the total American investment of $550 million in Chile.

Chile may perhaps be acting rashly but now has clearly established the primacy of its own interests in copper working. The new attitude may perhaps be summed up by comparing the speech delivered to the Peruvian Mining Engineers Convention at the end of 1969 by the Chilean minister, Alejandro Hales, with Senator Tomic's remarks eight years earlier.[5]

We do not want, nor shall we accept, a repetition of the sacrifices of the past, in which our nations were limited as to their legitimate income, while the purchasing countries returned their industrial products, but always at higher prices . . . We are not robot suppliers in a world which needs our raw materials, nor simple tax collectors; we are an essential part of the copper industry of the world. No decision may be made without our own decision. Our attitude is positive and clear. Nobody has to fear us, but nobody can do without us.

Africa

The most important single copper district in the world straddles the Zambia–Congo border. Over 18 per cent world 1969 output came from its mines. The surface is generally flat – part of the African plateau at about 4,000 feet – with a landscape diversified by a few residual hills, an open forest of acacia trees and shallow grassy hollows, known as 'dambos' which become swamps at flood time. Beneath this is a broad mineralized belt 300 miles by twenty to forty miles, the long axis running north-west to south-east across an international boundary agreed long before the minerals were discovered.

In 1911 production of good quality oxide ores was begun in Katanga. Already the deposit had been traced over the border into Northern Rhodesia but the copper content there was only 3 to 5 per cent as opposed to 7 per cent in the Congo. At this time the north of Rhodesia was a sparsely populated cattle country with no rail access. Development of the first Northern Rhodesian mine was begun in 1903, the Rhodesia Railway reached the Congo boundary in 1909 and the first Northern Rhodesian copper shipments were made in 1913. For long attention was centred on the search for oxide ores of grades comparable with those of Katanga. Thick laterite cover and the rareness of outcrops made geological exploration difficult. In 1923 search was concentrated in the hands of a smaller, more powerful group and systematic survey proved that at a depth of about 100 feet the deposits entered a sulphide zone. This radically changed the economic prospect, for sulphide ores of 3 to 5 per cent were profitable where oxide ores of similar grades were almost valueless. Development work began in 1929 with copper prices high and in mid 1931 Roan Antelope became the first big producer. By this time prices were slipping towards the trough of the Great Depression – from a 1929 maximum on the New York market of 21.25 cents a lb. to a 1933 minimum of 4.75 cents. There was a sharp Rhodesian slowdown, mine employment and associated construction employment falling between 1930 and 1932 from 19,000 to 8,000 and from 30,000 to 7,000 respectively. Luckily at this stage permanent mine settlements had not yet been established on the Copper Belt so that the displaced labour could be reabsorbed reasonably easily by the tribal system. Because development work was already far advanced, Rhodesian producers would not join in the output restriction scheme, and by 1935 had acquired about 10 per cent of non-communist world production. By the late 1930s Northern Rhodesia was well ahead of the Congo.

The mineral wealth of Katanga was extracted for sixty years by the Belgian Company Union Minière du Haut Katanga. Union Minière made immense contributions to the Belgian economy, partly by providing raw materials, partly through its profits paid both to big finance houses and to private shareholders, said to number 120,000 both in Belgium and in France. Within the Congo by the mid sixties it had an African labour force of 20,000, and 2,000 Belgian employees. It was a paternalistic influence and a vital financial prop both to the provincial and the Congo government. In 1951, when the prospect of Congo independence seemed something to be seriously entertained perhaps a century hence, the Chairman of Union Minière summed up the symbiosis of his company and the colonial government, and the prevailing paternalistic philosophy.

The Government of the Colony and the Comité Spécial du Katanga, to whose confidence we pay tribute, will be further convinced from year to year that our main concern has been the general interest of the country and the moral and material fate of the Congolese population, which is under our care and which leads a happy family life in our social centres.[6]

Congolese independence in 1960 was quickly followed by chaos, but mineral output was kept up and mutual support kept alive both the Union Minière and the Katanga secessionist regime of Moise Tshombe. Later, after the reunification of the country, taxes and duties on Union Minière exports provided some 50 per cent of the Congo budget and 70 per cent of her foreign exchange. The basis for all this is found in seven open pits and three underground mines. Katanga ore is fairly rich by present day standards but costs of production are high because of high charges for power and transport, low recovery rates, heavy taxation and costly labour.

At the end of 1966 the state took over all the Congolese operations of Union Minière. However it appointed a Union subsidiary technical adviser to the new national mineral agency Gécomin (Générale Congolaise des Minérals), in return for payments of about £10 million annually. For contrast, in 1951, with output less than two thirds as great, £11 million was available for distribution to Union Minière shareholders. Compensation for loss of assets was requested but settlement was not reached until 1969. By the terms of the 1969 agreement the Congo will pay Union Minière 6 per cent of its sales revenue for fifteen years – 1 per cent for technical

131

assistance and 5 per cent for compensation. After that payment will be no more than one per cent of revenue. The prospect for the European shareholder is, however, brightening, for Gécomin plans to raise its output from 360,000 metric tons in 1969 to 450,000 in 1975, and 560,000 in 1980. In 1969 the mining group, Lonhro, which has big ambitions for a share of Congo mineral development, was speaking of even bigger developments involving a doubling of copper production in six or seven years. Other interests too are coming in. In collaboration with the Congo government, Nippon Mining is to open a 50,000 ton per annum mine at Musoshi south of Lubumbashi. The model of 'Congolization' provided a guide for newly independent Zambia.

Table 20: The leading copper producers in the capitalist world 1929–69 (million tons mine output)

	1929	1938	1948	1954	1965	1969
United States	0.89	0.50	0.74	0.75	1.21	1.36
Chile	0.31	0.35	0.44	0.36	0.59	0.67
Congo	0.13	0.12	0.15	0.22	0.28	0.35
Canada	0.11	0.25	0.21	0.27	0.45	0.50
Zambia	0.01	0.21	0.22	0.39	0.68	0.70

Source: British Metal Corporation, Review of Non-Ferrous Metals.

In the thirties the Copper Belt established itself as a low cost copper producer and steadily increased its share of world business. By 1938 Northern Rhodesia was fourth ranking producer, by 1948 third, and in 1954 it pushed ahead of Chile to become second. In 1970 Zambia was expected to produce 750,000 tons of copper. The cave-in at Mufulira in September 1970 helped to reduce production to well below this target but Zambia remained ahead of Chile. Two groups, Anglo-American and Rhodesian Selection Trust (now renamed Roan Selection Trust), dominated its growth.

The relationship of open-pit and shaft mining on the Copper Belt has been interesting. Anglo-American shaft operations at Nchanga were replaced in the mid 1950s by open-pit working. The change proved highly successful with very low operating costs. However, as open pits become deeper so stripping of waste becomes uneconomic. At Chambishi, the Roan Selection Trust has decided to test the feasibility of underground mining. In this case a critical factor concerns the suitability of ground conditions for the large openings necessary to make a success of mechanical mining underground. Water conditions are vital and their significance has recently been tragically brought home.

Heavy watering has been a problem and an important cost factor in Copper Belt operations. In 1957 Anglo-American's Kansanshi mine was completely closed when an underground stream broke in. Within less than a month almost all the 500 to 600 African employees and their families had left this rather remote community to find new jobs in the main developed area. The mine was not reopened until 1970. After £18 million had been spent on five years of development work, the big Bancroft mine came into production in 1957. Refractory ores, falling prices and water problems quickly closed it. By 1970 the pumping rate at Bancroft was 72 million gallons a day but it was to be increased to 130 million in the hope of lowering the water level sufficiently to open up new lower seams. In the late summer of 1970 came the tragic cave-in at Mufulira mine which put it out of production and caused the death of eighty-nine men.

17. *Mufulira Mine, Zambia.*
This Copper Belt mine, one of the largest in the world, suffered a disastrous
'cave-in' in the late summer of 1970. Note the isolated setting on the flat and largely
unpopulated plateau.

Mineral Resources

In recent years an important breakthrough has been made in the technology for dealing with formerly refractory, low grade ores. Anglo-American's T.O.R.C.O. (Treatment of Refractory Copper Ores) process makes it possible to work lower grades and so lengthens out the lives of existing mines. Similarly, it brings formerly sub-marginal prospects above the line. Two beneficiaries of the process have been the flooded Kansanshi mine and Bwana Mkubwa, the 1903 precursor of all Copper Belt development, where operations ceased in 1931 only to be restarted in December 1969.

Along with increases in production and technical changes have gone major changes in the organizational structure of Zambian mining. Cecil Rhodes vested the royalty rights from Northern Rhodesian mineral working in the British South African Company. Gradually this concern was brought under the control of local governments – for instance, by a 1948 agreement the company committed itself to turn over 20 per cent of its earnings to the Northern Rhodesian government and to pay full local taxes on the rest. In 1964 these royalty rights were transferred to Zambia. A United Nations report on the Zambian economy in the same year recommended direct government participation in mining. It took almost five years for Zambia to move decisively along these lines. However in the meantime some important changes were introduced. In 1966 an export duty of 40 per cent was imposed, and in 1968 a new body was established to control the country's processing and marketing policy. Copper provided 92 per cent of Zambia's export earnings in 1967 and royalties and export taxes in 1967–8 supplied an estimated K104 million out of the total government revenue of K297 million.* In August 1969 President Kaunda announced the Zambian government's decision to obtain a controlling interest in the mining concerns and by November agreement had been reached between the two companies and Indeco (the Industrial Development Corporation). Zambia acquired a 51 per cent share in their equity, to be paid for in instalments over twelve years in the case of Anglo-American and eight years for R.S.T., a shorter period because the R.S.T. mines have been rather

* 1.7 Kwachas = £1.

134

more profitable in relation to book value. A new mineral tax replaces both the old royalty system and the export taxes which were introduced in 1966. The old royalties were related to the sale price for copper on the London Metal Exchange and therefore badly hit the low profit margin mines. The new mineral tax will be equal to 51 per cent of net profit and the companies will pay regular company income tax on the balance, or a total tax payment of 73 per cent. Such a system encourages profitability and has been generally welcomed by Anglo-American and R.S.T. The Zambian government has lifted its former restrictions on dividend payment outside Zambia. The new arrangements came into force in January 1970, R.S.T. and Zambian Anglo-American providing management and acting as sales agents under contracts running for a minimum of ten years.

Finally, the Zambian government has foreclosed on the exclusive mineral and prospecting rights of Anglo-American and R.S.T., granted to them in perpetuity by the British South African Company. They have been offered to newcomers. Announcing these and other economic changes to his United National Independence Party in Lusaka in August 1969 President Kaunda set the tone for the triumphant new economic nationalism. 'Cecil Rhodes and his "in perpetuity for ever and ever" is now buried and, I hope and pray, never to rise again in this part of Africa.'

External forces set off by Rhodesian U.D.I. in 1965 have affected Zambian copper development. Copper deliveries and coal supplies over Rhodesian Railways were upset. The Zambian government refused to permit free movement across the boundary, requiring that each freight car coming from Rhodesia should be matched by one travelling in the opposite direction. A year after U.D.I. only half the coal needed was arriving and production of refined copper had been cut by 33 per cent. Development went ahead rapidly at the new Nkandabwe coalfield near Lake Kariba in 1966. Initial transport problems were overcome by the purchase of dumper trucks, and 160,000 tons of coal were delivered in the first half of 1967. In 1968 the road to Batoka railway siding was improved. However, even from Batoka the journey to Ndola in the Copper Belt is 300 miles, and the Nkandabwe coal has a high ash content.

135

Mineral Resources

By 1967 work was underway on the development of an open pit in the better coals of Maamba, formerly Siankandoba. Nkandabwe was closed at the end of 1969 as a worked-out mine. The transport problems for Zambian copper since 1965 are examined in chapter 4 (p. 80).

Under Indeco's overall control, both Zambian copper producers are making considerable increases in output. In 1968 negotiations with two Japanese firms secured a loan of $42 million for Zambian Anglo-American. In addition it will receive from Japan $28 million of machinery and equipment on a deferred payments basis. In return it will supply 100,000 tons of copper each year for ten years. Increases of output in one developing country after another create problems in the world market for copper.

Demand, Supply and Copper prices

Between 1961 and 1966 world copper consumption is estimated to have risen by 5.6 per cent per annum. Supply to meet this demand comes very largely from new mine capacity, though scrap copper is an important source of supply in the advanced industrial countries. In the United States, for instance, the supply of new refined copper from domestic and foreign sources was 1.3 million tons 1968 and from scrap 0.45 million tons.

Growth in consumption is the result of rising living standards in advanced industrial countries; it has received a great boost in recent years from Japanese economic growth. Between 1956 and 1966 Japanese production of copper from home mines went up from 83,000 to 111,000 tons but by the latter date her production of electrolytic copper was already 405,000 tons. By 1969 Japanese output was 120,000 tons but her refineries produced 629,000 tons. In the early 1970s their output will reach one million tons.

While demand pressed upwards, supply was held back for several years by interruptions to production – major Chilean strikes and Rhodesian U.D.I. late in 1965, and in 1966 Chilean strikes again and the Rhodesian–Zambian railway dispute. In 1967 and 1968 60,000 U.S. copper industry workers were idle for eight and a half

Table 21: Production of copper ore, smelted and refined copper and copper consumption 1967 (thousand tons)

	Mine output	Smelter output	Refined copper	Consumption of refined copper
Chile	660	630	353	17
Peru	186	165	35	?
Congo	321	321	161	?
Zambia	663	638	535	?
United States	865	846	1396	1755
West Germany	1	121	382	515
Japan	117	386	470	616
United Kingdom	nil	?	179	514

Source: Quinn's Metal Handbook.

months with the loss of 850,000 tons of refined copper, which required stock pile releases to make up some of the shortfall. As a result prices rose. It is important, however, to qualify this statement for there have been three major prices for copper in non-communist markets.

The most volatile prices are those of the London Metal Exchange. These are spot quotations, that is short term prices which change twice daily, so closely reflecting the supply/demand situation. Averaged out, these show considerable variations, but in the late sixties there was an upward trend of prices – £443 a ton in 1965, £537 a ton in 1966, £404 a ton in 1967, £501 a ton in 1968 and £603 a ton in 1969. In late March 1970 London Metal Exchange prices reached £755 a ton.

The second price dates from the late 1950s when prices began to fall. The main producers of the Third World decided to try for greater stability than L.M.E. prices give. Until late 1965 the Producers' Posted Prices which resulted from their efforts were kept lower and steadier than L.M.E. prices. After that, however, they closely mirrored the L.M.E. variations. Finally U.S. producers set their own domestic prices, which are much more stable than L.M.E. ones. As the latter moved strongly upward in the late 1960s, U.S. prices have remained well below them even though their rate of increase has been much more rapid in the last three years.

Table 22: London Metal Exchange and U.S. Producer Prices for Copper
1960–70 (cent/pound)

	L.M.E.	U.S.
1960	30.7	32.3
1961	28.7	30.3
1962	29.2	31.0
1963	29.3	31.0
1964	43.9	32.3
1965	58.5	35.4
1966	69.4	36.0
1967	51.1	38.1
1968	56.3	41.2
1969	66.5	47.5
1970	64.0	58.2

Source: Mining Annual Review, 1971.

Copper has to face the competition of substitute materials such as stainless steel, plastic and above all aluminium. A metal fabricator will not lightly change to a new material for this will involve

Table 23: Recent growth in world consumption of main non-ferrous metals (per cent per annum)

	1965–67 average	1968	1969	1970–75 or 1980 (estimate)
Copper	1.0	7.6	9.4	4.5
Lead	2.5	6.7	5.7	3.75
Zinc	2.7	8.5	9.5	4.25
Tin	0.5	3.1	3.7	2.25
Aluminium*	?	6.8	10.9	9–10.0

* Production.
Source: British Metal Corporation, Review of Non-Ferrous Metals.

new techniques and possibly new tools, but once the change occurs reversion to the old material is unlikely for the same reasons. However, it has been suggested in some quarters that the wide variation in copper prices has paradoxically checked the substitution process, for the fabricator cannot be sure that he will not lose in the change. As will be seen the events of 1970 must have reinforced such caution.

A few years ago it was widely maintained that if copper ran

138

above £240 a ton a cataclysmic switch to aluminium would follow. By March 1970 prices were more than three times as high and there was no sign of a collapse in demand. There has, however, been a substantial transfer to the competing material. Aluminium has an electrical conductivity only 63 per cent that of copper, but the material is both lighter and cheaper, and both its supply and its price have been more dependable. In the last eight months of 1968 U.S. domestic copper prices were 42 cents a pound, by early 1970

Table 24: Average Prices of Copper and Aluminium in the U.S.A. (*cents per lb*)

	1939	1950	1961	1968
Electrolytic Copper, f.o.b. U.S. refineries	11.07	21.23	30.14	41.85
Unalloyed Primary Aluminium (99.5 per cent pure)	20.0	17.69	25.3	26.0 (June)

56 cents. The price for aluminium rose much more slowly by two cents a pound in 1969, as opposed to ten cents for copper, to a figure of 29 cents by spring 1970.

Outside the U.S.A. reduction in copper demand in 1966 and 1967 as a result of substitution was estimated at 440,000 tons, equal to 14 per cent of world mine output in 1967 outside the U.S.A. and the communist bloc.[7] In Britain, electrical industries account for about 40 per cent of copper consumption but there has been a good deal of transfer to aluminium. In 1966 British Insulated Callendar Cables announced that by 1968 one third of its wire and cable business would have been changed from copper to aluminium, and by 1970 half. A noteworthy sign of this changing emphasis was B.I.C.C.'s share in the Holyhead aluminium smelter. Previously it had had interests in copper expansion. On the other hand it is easy to exaggerate the pace of substitution; 1970 reports suggest that in only a few United States consumption lines has the inroad been as much as 5 per cent a year, but the likelihood of substitution checks the freedom with which copper producers can raise prices, and it reduces potential demand. This is of great importance as copper production costs are rising rapidly and at the same time and for a variety of reasons supply is increasing rapidly too.

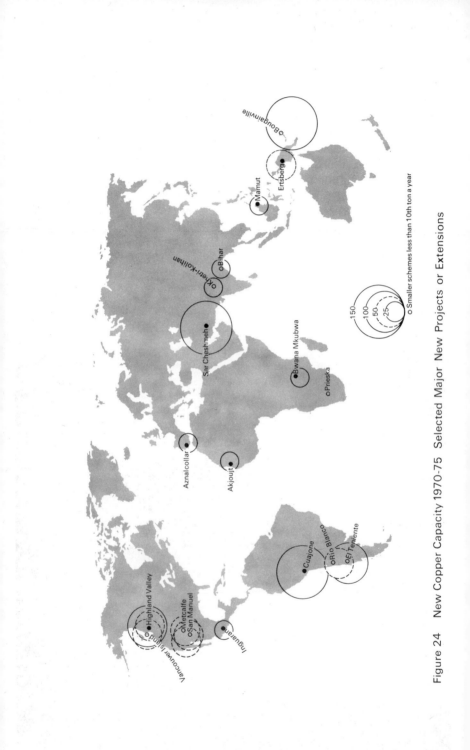

Figure 24   New Copper Capacity 1970-75   Selected Major New Projects or Extensions

The Japanese have been extremely active in developing new mines. By 1969–70 a consortium of six Japanese concerns were involved in big prospecting and copper mine development in Katanga, Japan is concerned in new developments at Kinsenda in the Congo, and leading firms are to buy copper from the Ertsberg mine in West Irian, Indonesia, have underwritten the huge Bougainville project, have interests in the Chaca Valley, Colombia, and in the Chapi Mine and now the huge Cuajone project in Peru. Other interests of Japanese concerns either directly or in loan finance are in British Columbia, in the Kerman project of south central Iran and in Mount Isa, Queensland.

18. *Bougainville Copper mine under development.*
Full-scale commercial production from the 900 million tons of 0.48 per cent copper deposits will begin in 1972. In this case the environment is Tropical Rain Forest.

Elsewhere too production is being increased on a big scale. This is partly because of the desire of developing countries, now rapidly establishing their influence over foreign mining companies, to increase their revenue from copper exports. The case of Chile has been considered above. In 1969 56 per cent of Peru's foreign exchange earnings came from copper, and in the summer of that year the government took back long unused concessions and stung the companies into big investments leading to its aim of tripling copper production in the next ten years to 600,000 tons.

In the world as a whole to late 1973 copper mining capacity is likely to increase by 37 per cent or 1.9 million tons – equal to the whole output of North America. This is leading to short-term surpluses – capacity until late 1973 increasing at 7.1 per cent a year but demand going up annually by no more than four to five per cent. According to the Anaconda's chairman, by 1975 there could theoretically be surplus world capacity of 0.85 million tons. However strikes such as those in Chile in the autumn of 1970 and possibly at the time of the U.S. wage negotiations in 1972, and unpredictable breakdowns in supply, such as the Mufulira cave-in, will reduce this theoretical surplus.[8]

In the course of 1970, as supply began to push beyond demand, prices tumbled from an L.M.E. price of £755 at the end of March to £450 in October and to £421 by early December. This collapse was of great significance to countries heavily dependent on income from copper exports. In June 1967 representatives of Chile, Zambia, the Congo and Peru had met in Lusaka to attempt to integrate their programmes of copper development. The result was a permanent joint council for communication and for coordination of interests. The Intergovernmental Council of Copper Exporting Countries (C.I.P.E.C.), made up of representatives from the four, was naturally disturbed by the price falls. Suggestions that they might try to keep up prices by restricting output are ruled out by their commitment to increase production. Price fixing or stockpiling are beyond their power. By December 1970 they had still not revealed how they would attempt to bolster prices. Their wish to do so is likely to be one factor in changing the world pattern of copper production and trade. Early in 1971 L.M.E. prices fell to little above £400 a ton but had strengthened to £500 by late March.

Price support, strong interference with the freedom of expatriate concerns and inflationary pressures on costs seem likely to adversely affect the Latin American and Central African producers. Japan has shown the way with its large long-term loans in exchange for assured supplies. Some of her agreements, it is true, are with C.I.P.E.C. countries, but the biggest developments are elsewhere. The big European and North American companies which have historically gained greatly from low African and South American costs will probably transfer attention to big operations in areas where stability will be more assured. In 1970 Sir Ronald Prain of R.S.T. International suggested that within five years the C.I.P.E.C. share of world copper exports will fall from 72.1 per cent to 61.4 per cent.[9]

Meanwhile costs of production are going up everywhere so that by 1969–70 it was reckoned that a price of £500 a ton for copper was normally necessary to make the new mines profitable. In 1970 the agreement to bring the Cuajone ore body into production in 1976 provided for development costs of $2,535 per ton of annual output. Late in 1970 the President of Zambian Anglo-American estimated that the capital cost of a five-year, 25 per cent stepping up of Zambia's production, taking into account the maintenance of production from existing mines, would be not less than $2,016 a ton. Sir Ronald Prain has suggested that over the last decade the cost of bringing in new capacity per ton of refined copper had risen from $1,700 to $2,800.

In short, copper producers are caught between the devil and the deep. Prices must be high to give a return on investment, but if they are too high substitution will go on more rapidly. New mines and trade lines are emerging, but the old producers are trying to boost output for wholly understandable development income reasons yet apparently without regard to the effect on prices. In the long term there is every indication that copper demand will keep on growing, but the short term outlook is extremely uncertain. Copper shares would seem to be suitable for long term capital growth but to be avoided by those who look for quick returns.

7. International Trade and Price Problems in Minerals – the Cases of Nickel and Tin

Since the mid nineteenth century base metals have joined the precious metals as major items in world trade; first Europe, later the United States, and finally Japan becoming major purchasers from the Third World. The imbalance between world production and consumption has grown. In order to protect the investments of its nationals abroad, to have some control over the cost of its basic raw materials for manufacturing, and sometimes to secure strategic materials, the government of each advanced industrial country is keenly interested in the course of world mineral production and trade. In their turn, as more initiative is acquired and their colonialism or economic subservience is shaken off, the developing countries are vitally concerned with the relationship of their production to the consumption of the developed world. They are keen to ensure the most favourable relationship between the price which they pay for imported manufactures and those received for raw material exports. As minerals are wasting assets each country exporting its natural wealth in this form wishes to secure permanent benefit in the form of investment in other than primary industry. International economics and the politics which give them colour play a major role in the world's mineral industries.

Early in the twentieth century the British and Russians manoeuvred for spheres of influence in Persia when oil supplies for the Royal Navy were at stake. The German drive to the Ukraine in 1941 was in part the result of a desire to control Caucasian oilfields, and also to secure the manganese field of Nikopol in the west of the industrialized zone. The situation of U.S. steel-makers highlights the strategic factor, for of all the various main alloy metals, the U.S. has adequate home supplies only of molybdenum.

Dependence on outside supplies can be eased but rarely removed.

Scrap metal recovery is important with many metals, particularly in the case of steel, but also for copper, aluminium, zinc and lead. The effect of scrap on supply and price situations is limited by its relatively small and inelastic supply, especially when demand is rising steeply. Other possibilities are economy in use and employment of substitutes, and yet another, and over the last thirty years a very important one, is stockpiling. In the late thirties Germany built up stockpiles of most strategic materials, especially of oil and manganese. Japan at the same time was accumulating stocks of steel scrap, largely by purchase from the U.S.A. More recently the U.S. government and, on a much smaller scale, other western governments, have operated stockpiles of most minerals for strategic purposes – strategic both in the military sense and also to ensure manufacturers against interruption of normal commercial supplies. A final effect of the danger of supply breakdown is to encourage the opening of new mines, either overseas or, perhaps, lower grade bodies at home. Some of these aspects may be illustrated by the recent history of nickel production.

By the late 1880s it was shown that nickel provided an essential toughness to armour plate. Subsequently strong, tough, nickel steels were also used largely in ordnance manufacture. Later, happier uses were found in the critical parts of vehicles, constructional material and so on. Recognition that it gives improved strength, stability and corrosion resistance at high temperatures has made nickel a vital factor in industrial turbines and jet engines. Another major use is in the manufacture of stainless steel, which in 1967 took 37 per cent of the nickel consumption in the non-communist world. World production and consumption of nickel reached its earliest peaks in the two world wars – 52,000 tons in 1918 and 169,000 in 1943. By 1968 consumption in the non-communist world alone was 360,000 tons and 1970 forecasts have looked to an eight to nine per cent annual increase in consumption to 650,000 tons by 1975.

There have been relatively few sources of supply. Until 1905 growing consumption was met from the nickel silicate ores of New Caledonia, but, after that, the sulphide ores of Sudbury, Ontario, made usable as a result of a fortunate breakthrough in nickel

145

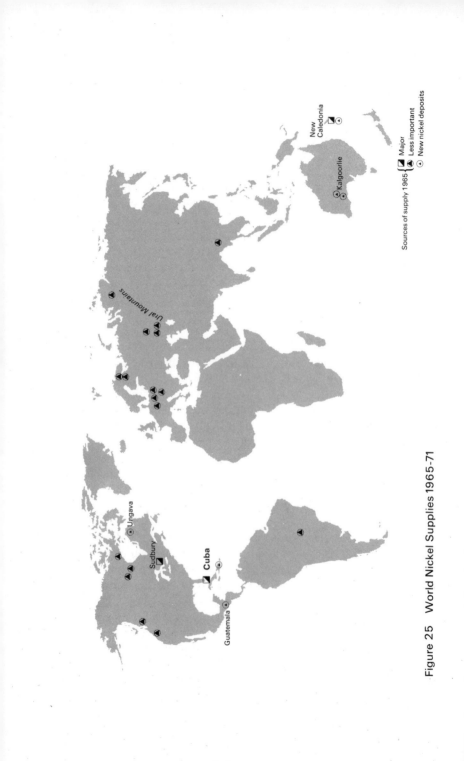

Figure 25 World Nickel Supplies 1965-71

refining technology, dominated world production. International Nickel at Sudbury, Le Nickel of New Caledonia and, after 1929, Falconbridge Nickel Union, also Canadian, dominated production. In the Second World War, as demand spiralled and Pacific sources of supply became uncertain, the U.S. financed the opening of nickel deposits in north-east Cuba. After the war the Cuban operation followed an erratic course – closure in 1947, re-opening in the Korean War, run-down in the late fifties and then, following Fidel Castro's victory, Cuban ore was lost to the western world. Demand was now growing rapidly, averaging 18 per cent annually from 1963 to 1966. By the late sixties 70 per cent of the non-communist world's nickel came from Canadian mines, 20 per cent from New Caledonia, while the U.S. and South Africa produced 4 per cent and 2 per cent respectively. U.S.S.R. output is about 40 per cent that of Canada.

By 1967 there were new projects in Canada, in New Caledonia, where output is increasingly geared to the needs of Japan, in Minnesota and Guatemala. However, shortages were developing, producer's inventories were declining, and between mid 1965 and early 1967 the U.S. government ran down its nickel stockpile by three quarters. Believing that it was in the long-term interests of the business, producers tried to check the pace of price rises, but in 1969 a strike of Inco and Falconbridge employees lasting over eighteen weeks upset the whole supply situation and prices rocketed to unprecedented levels.

Already shortages had gradually pushed up the official price of Canadian producers – from £622 a ton in 1962 to £986 by 1968. With the new shortages free market prices moved rapidly out of line. By October 1969 they reached the astounding figure of £7,000 a ton. In November 1969 the strike was settled, but with terms which will push up costs and therefore the normal prices of the Canadian firms.

Demand projections to 1975 vary widely, but all agree on a very large increase. Taken with the inevitable higher prices, this has brought a great surge in nickel search and development work. Even before the 1969 strike it was estimated that £520 million would be spent between 1969 and 1973 in a non-communist world expansion

19. *Clarabelle nickel-copper concentrator, Sudbury, Ontario.*
This new crushing and concentrating plant costing $80 million will process
35,000 tons ore daily.

of as much as 48 per cent. Major extensions are planned in Canada
and even bigger ones in New Caledonia. Lack of capital, know-how
and loss of the U.S. market have hampered Cuba over the last
decade but she now plans to increase nickel exports as one means

of reducing her trade deficit. The most spectacular developments have been in Australia.

Australia produced its first nickel in 1967, but by 1969 was in the throes of a prospecting and development fever. By mid summer there were 120 claim holders in an area centred on Kalgoorlie and

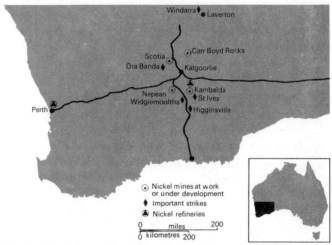

Figure 26   The West Australian Nickel Belt 1970

extending a hundred miles north–south and sixty miles east–west. Later in the year the wild speculative rush associated with the nickel discoveries of the Poseidon group caused its share values to appreciate two hundredfold in four months. By late 1969 Western Mining and the Metals Exploration Company had two nickel mines in the Kambalda area. In some cases the Australian ore runs up to as much as 4, 5 or even 6 per cent nickel and more commonly is of 3 per cent quality. Western Mining's ore is of 3.9 per cent nickel. This compares well with the Inco ore which was reckoned to average 1.5 per cent. Yet Inco has long-established operations, the advantages of scale and of the joint production of large tonnages of copper. Even though the Sudbury area is by no means temperate, most of the infrastructure has been provided long ago, transport facilities are good and the U.S. market is highly accessible. Finally,

149

20. *Kambalda nickel mine, Western Australia.*
This first Australian nickel mine, 40 miles south of Kalgoorlie, was opened to
work deposits of up to 100 million tons of 4 per cent nickel ore. By April 1969
2,500 lived in Kambalda company town and by 1972 population was expected to
rise to 6,000.

150

in the autumn of 1970 Inco revealed that it is even more competitive by announcing that its nickel reserves grade much higher than had been believed – 2.67 per cent.

There is considerable dispute about the supply/demand situation by the mid seventies. It is widely agreed that consumption will be about 650,000 metric tons by 1975 but whereas Inco and Falconbridge estimate non-communist production at a maximum of 715,000 to 725,000 tons, the president of Le Nickel has suggested a figure of 830,000 tons. Even allowing for the necessity of operating below full-rated capacity, there seems a possibility by that time of over-supply and price falls.

The Implications of Mineral Price Variations

Mineral price variations are naturally of importance to both producers and consumers. In the case of the former, they affect the import bill, cost of manufactures, and therefore the price of exports. However, the more elaborate the processing the smaller the effect of an increase in the price of concentrate or metal. Moreover, as their economies are complex and their per capita wealth high, the advanced nations are less substantially affected by price rises for mineral imports than the strength of their complaints would suggest. Producers suffer much more.

If the price of their mineral exports rises too high, then a switch to substitute products is possible; if they fall low then their export revenue and vital development funds are seriously depleted. In many cases one or two minerals, along with similarly price-volatile agricultural products, dominate their export business. Price gyrations and their generally unfavourable trends as compared with price levels for imported manufactures constitute the intractable commodity price problem of the Third World.

At the beginning of the century, Alfred Marshall looked ahead to a time when the initiative in trade and prices would lie with the raw material producing and exporting countries, but there is no sign of this situation yet. There has been a long-term trend for primary product prices to rise less than those for manufactures, a long-term

151

swing of the terms of trade against the underdeveloped world. Superimposed on this secular trend are wide short-term variations. These reflect the fact that demand for raw materials is price inelastic, that is, generally speaking, it varies greatly with the state of the economy of the advanced industrial countries but is not much affected by raw material prices. A substantial fall will not significantly increase the demand for nickel or zinc or tin. On the other hand elasticity of supply is low, mines take a long time to develop, and once opened represent such a substantial standing charge that the producer is reluctant to restrict output. With the new nationalistic drive to increase mine output and mineral income, this tendency for supply to be inelastic downwards will increase. A fall in the price for a mineral may deprive an underdeveloped country of more income than it receives in foreign aid – as was certainly the case with the mineral price collapse in the 1956–8 period. Such a fall brings pressure on the balance of payments and requires remedial action. This may take a variety of forms, all of them inimical to economic advancement and to the current standard of living – deflation, physical restriction of imports or a depreciation of the currency. The primary producing area of the country will be especially severely hit. It is widely recognized that efforts to avoid these dangerous price variations are to be preferred to piecemeal remedial action. Achievement so far has been small.

Theoretically price variations may be ironed out to some extent by the use of a buffer stock. A mineral – or an agricultural commodity – is bought when prices start to sag and stocks are sold to check too rapid a rise. The effectiveness of a buffer stock is limited by the extent of cooperation between producers and consumers, by the finance which the buffer stock manager has at his disposal and by the wisdom with which he exercises his powers to buy and sell. A second device is to restrict output. A danger with restriction schemes is that if prices are kept high the consumers may shift to alternative sources or to another material. There may be temptation for a producer to defect from the restriction scheme in order to sell a greater tonnage.[1] Moreover 'Chileanization', 'Zambianization' and so on show clearly that the developing countries are less and less willing to restrict output. The advanced industrial country retains the initiative, being able to divert its purchases to more amenable producers or perhaps to boost higher cost production at home or in the more stable parts of the developed world. The recent history of lead and zinc mining illustrates this problem.

Prices for lead and zinc have been moving up in the late 1960s but the situation over a rather longer time span, covering the last dozen years, shows how a favoured advanced industrial country can export depression in base metals and yet keep boom at home. In the late fifties, with the decline in non-ferrous metal prices, employment in American lead and zinc mines and smelters also fell sharply – by 50 per cent between 1952 and spring 1960. The U.S. industry pressed for protective tariffs, but instead was given quota protection in 1958 – imports of the two metals were to be restricted to 80 per cent of the 1953–7 level. It was reckoned that this would exclude perhaps 152,000 metric tons of lead and 248,000 tons of zinc. U.S. zinc ore output especially rose rapidly after this and by 1970 was 50 per cent higher than in 1958.

It is true that in this case other advanced economies were very hard hit by U.S. policy, but so too were Mexico and Peru. In addition to the diversion of demand to U.S. mines the attempts of other producers to find alternative outlets led to price falls in markets outside the U.S.A. By the summer of 1960, U.S. zinc prices were more than one sixth higher than those on the London Metal Exchange, and for lead one quarter more. Outsiders had been forced to bear the brunt of the problem of adjusting supply to demand and the U.S. market had been partially insulated. Meanwhile the minority report of the U.S. Tariff Commission suggested that the lead–zinc quota was insufficient and that a tariff was necessary '. . . if a reasonable opportunity is to be provided for U.S. lead and zinc mines to operate on a sound and stable basis'.[2] There are limits to the degree of insulation which can be achieved. The U.S. cannot supply all its needs of most base metals. It consumes about 1.2 million tons of zinc a year but mines under 500,000 tons. Most Canadian output goes to U.S. consumers. Moreover, as with copper, major U.S. producers are involved in very big operations overseas. The St Joseph Lead Company, said to be the world's largest zinc producer, has interests in Peru, and Anaconda, Asarco

153

and Texas Gulf have substantial investment in Canadian zinc production.

The attempts to shield an advanced industrial economy at the expense of the primary producers is by no means confined to the U.S.A. In the last few years the long derelict lead and zinc mines of Derbyshire and the northern Pennines have been re-evaluated, tin mining expansion is actively underway in Cornwall and prospects for gold and copper awaken the fears of those anxious to preserve the beauty of the Mawddach estuary. Canadian and now British interest is involved in base-metal development in Eire where zinc output went up from 22,000 tons in 1966 to 96,000 tons in 1969. Other base-metal operations are developing in County Tipperary and in the Wicklow Mountains. More modestly, German mines stepped up zinc production to 137,000 tons in 1969. In the case of no other metallic minerals have the variations in prices, the problems of regulating business and the effect on developing economies been more acute than in the case of tin, even though advanced industrial countries have very little share in its production.

Tin

Consumption of tin has grown much less rapidly than that of most other metals. In 1969 world production was a little under two and a half times the level of 1905. Lead output was three and a half times and copper eight times the 1905 level. There are few major tin producers though a large number of smaller ones and, in the non-communist world, they are all underdeveloped countries; North America effectively mines none at all. There are considerable variations in costs of production. Attempts to regulate the tin market have been made over a period of more than forty years but, although short-term success has been won, time and again eventually each regulation scheme has foundered. Prices have followed an uncertain, even wild course, in the past twenty-five years since 1945 ranging from a low of £301.5 a long ton on the London Metal Exchange to a high of £1,715.

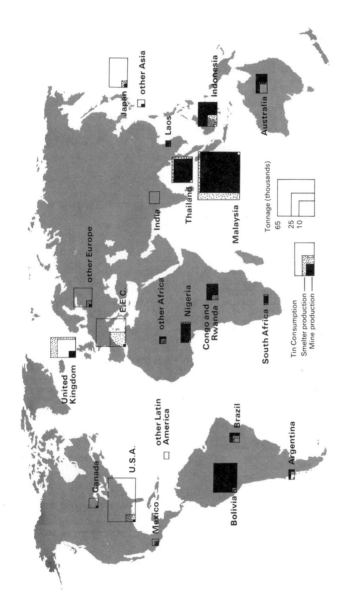

other Asia

Japan

Laos

Indonesia

Australia

India

Thailand

Malaysia

other Europe

E.E.C.

other Africa

Nigeria

Congo and
Rwanda

South Africa

Tonnage (thousands)

65

25
10

Tin Consumption
Smelter production
Mine production

United
Kingdom

U.S.A.

other Latin
America

Canada

Mexico

Brazil

Argentina

Bolivia

Figure 27   Production of Tin and Tin Consumption 1970   Based on International Tin Council Statistical Bulletin

Mineral Resources

The tin industry is strangely unbalanced. Although production of ore is confined to relatively few underdeveloped countries, smelting is divided between a handful of producers and consumers, at first consideration a haphazard handful of consumers. Consumption of metal much more closely reflects the standard of living or, more accurately, as use is concentrated in tin plate production, the state and type of industrialization of the country concerned.

Table 25: Production of Tin Ore and Metal and Tin Consumption 1967 (thousand tons)

	Tin ore (tin in concentrates)	Smelter Production	Tin Consumption
Malaysia	72	76	N.A.
Bolivia	27	1	N.A.
Thailand	22	26	N.A.
Indonesia	14	1	0.1
Nigeria	9	9	N.A.
United Kingdom	1	23	17
Netherlands	–	14	4
United States	–	3	57
Japan	–	N.A.	20
Non-Communist World	172	N.A.	169

Source: *Metal Handbook.*

Tinplate provides an excellent example of the power of modern technology to reduce dependence on the products of the under-developed world and their price uncertainties. A rise of £100 a ton in the price of tin is said to increase the cost of producing an average 'tin can' by less than one fiftieth of a new penny. Even so tinplate producers have acted both to economize in tin use and to find alternative materials. Substitutes such as plastic, aluminium or 'tin-less' tin cans made of steel have made headway, and improved coating techniques have cut down the use of tin even in tinplate, with the replacement of old 'hot dip' tinning by the electrolytic process and with the production of 'thin tinplate' in which the coating is thicker on one side than the other. As a result of these developments, demand for tin from the British tinplate industry

156

fell from 1947 to 1961 even though consumption of preserved foods doubled. In the U.S.A. tinplate production went up between 1960 and 1967 from 5.7 to 6.3 million tons, but tin consumption by the tinplate industry fell by 11 per cent. While this process is to the satisfaction of the accountants of the big steel firms, the incomes of such variously poor countries as Thailand, Indonesia, Malaysia, Nigeria and Bolivia suffer. Twenty years ago tin production was equal to nickel production. Already, the non-communist world's nickel production is approaching twice the tin output of the whole world.

The earth has only a few tin-rich areas. Historically by far the most important was the Hercynian fold belt of western and central Europe, and including Brittany, Cornwall, Saxony and Bohemia. In the eighteenth century the last three seem to have dominated world production. After that, for three quarters of a century Cornish mines were pre-eminent, English production of tin-in-concentrates* rising from 2,500 tons in 1801 to an average of 10,000 tons after 1860. In the early decades of the nineteenth century the Dutch East Indies island of Banka, long a small producer, became more important and later its neighbour, Billiton, joined the lists. After the middle of the century Malaya and Bolivia started production, but as late as 1875 the output of the former was only one fifth that of England and of the latter no more than one twentieth. Australia mined tin on a small scale but output did not grow spectacularly there after 1900 as it did in South-east Asia and Bolivia. Nigerian production began in the first decade of the twentieth century (see Table 26).

In smelting, Malaysia, Thailand and the United Kingdom are most important. The reasons for this pattern are to be found partly in technical, partly in historical circumstances. The prominence of the United Kingdom was related first to the importance of nineteenth-century Cornish production. As this dwindled British capital opened tin fields in the empire and throughout the developing world. British links with Bolivian mining have been close partly for technical reasons. Bolivian concentrates are particularly intractable and can best be smelted when mixed with ores from elsewhere.

*That is the metal content of the concentrates shipped from the mine.

157

Mineral Resources

Table 26: Mine output of Tin, 1905, 1939, 1969 (thousand tons metal)

	1905	1939	1969
Malaya	60	43	16
Indonesia	13	21	16
Thailand	?	13	21
Nigeria	—	7	9
Congo	—	7	7
Bolivia	12	25	30
Australia	5	3	8
Sino-Soviet bloc	?	c12	49
United Kingdom	4	2	1–2
World	94	149	229

Sources: Various.

Hence wide interests in tin mining, technical know-how and then later the inertia of established business connections explain the persistent prominence of the United Kingdom as Bolivia's chief smelter. In 1969 Bolivia produced 29,572 tons of tin-in-concentrates. United Kingdom smelters dealt with 23,819 tons of this, and 89 per cent of the concentrates imported to Britain were Bolivian. Again, largely for historical reasons, almost two thirds of Indonesia's tin is smelted in Malaysia. Thailand smelts her own ore and also smaller quantities of imported material. Practically all the ore mined in Nigeria is smelted there. The peculiar situation of the United States deserves fuller comment.

The U.S.A. consumes more than one third of the non-communist world's tin metal, smelts very little and produces almost no ore – production from western mines is about 100 tons a year but U.S. consumption ranges from 50,000 to 60,000 tons. Until the 1890s the few attempts to set up a domestic tinplate industry were ruined by Welsh competition. In 1891 the McKinley Tariff came into operation and a rapid build-up of the American tinplate capacity occurred behind a protective wall. Soon it had the world's biggest tinplate industry. Early in the twentieth century an attempt was made to build a tin smelter in New Jersey but the project was frustrated by British control of most of the tin ore in the empire and much of that outside. During the First World War several smelters were built near New York but British competition closed these in the twenties. In the Second World War the world's biggest smelter was built at Texas City near Houston on the U.S. Gulf Coast, to work largely on Bolivian ore. In spite of its size, its operating experience and its location within the world's leading market, post war competition has periodically closed it and it now concentrates on the secondary recovery of tin rather than on the smelting of concentrates.

The tin producers

Conditions in the leading producing countries vary widely. The extremes are set by Bolivia and Malaysia. In the former lode mining predominates, operations are conducted under extreme

Figure 28   South East Asian Mining 1970

difficulties of height and access. Malaysia has risen to undisputed world leadership on the basis of alluvial deposits worked in the most highly developed part of the country.

The tin deposits of the Malay peninsula were described by Arab writers in the ninth century. In the nineteenth century Chinese miners came in to open up innumerable small workings as output crept upwards in the Malay States and Siam. Their activities on the island of Puket in the Indian Ocean south-west of the Isthmus of Kra were described by a visiting European in the 1890s: 'The whole island is a gigantic tin mine. The granite of the hills is full of tin, the soil of the valleys is heavy with it. There is tin under the inland forests and tin beneath the sea. In search of tin the indefatigable Chinamen have transformed the scenery. The valleys have been turned inside out, the hills have been cut away; the sea has been undermined and the harbour has disappeared.'[3] Further south, in the Malay States, Britain intervened in a struggle for control of the mines between Thais, Malays and Chinese in the early seventies, and in 1882 the first European tin mining company to operate in Malaya was formed.

21. *Gravel pump tin working, Kampar, Malaysia.*
This mine, 102 miles north of Kuala Lumpur, uses water from a mountain stream in operations which require little capital and are therefore highly flexible. 972 gravel pump mines produced 55 per cent of Malaysia's tin, 1970.

22. *Landscape of gravel pump mining.*
Desolation of an area worked for tin. The wooden structure is a 'palong', down which the tin gravel and water mixture flows to sluices in which the heavy tin concentrates are separated out.

23. *Dredge mining, Malaysia.*
This mine is 6 miles north of Kuala Lumpur. Larger deposits and much bigger investments are required than with gravel pump mining. Dredges produced 32.4 per cent of Malaysia's tin in 1970.

Mineral Resources

Most Malayan tin comes from the western half of the peninsula. Though derived from primary deposits (lodes), it is mainly worked in placers. Perak and Selangor are the chief states, production being concentrated in the railway belt between Kuala Lumpur and Ipoh and especially in the small Kinta valley. In the early days there were a multitude of small labour intensive operations. The first steam engine and centrifugal pump was introduced in 1877 and dredging for the mass processing of the placers began in 1912. These developments caused a switch from very small operations. In the late 1920s, when tin prices were high, there was large investment in dredges, the number operating going up from twenty in 1920 to forty by 1925 and a hundred and five by 1929. Output increased and the labour force fell. By 1957, 37,000 workers produced as much ore as 200,000 workers in 1913. In the early years of capitalization much of the displaced labour, being immigrant, was sent back to China or to India.

Table 27: Malayan Tin: Output and Employment

	Output (thousand metric tons)	Labour force
1907	48	230,000
1931	53	79,000
1950	55	46,000
1969	73	46,000 (1970)

Unfortunately the dredges introduced a new inflexibility into mining. As they cost so much it was desirable to keep them operating even when prices were falling. As a result the smaller firms using gravel pumps became the marginal producers in the see-saw of prices. In the emergency of the 1950s, though the communists were gradually flushed out of the rain forest, prospecting almost ended for eight years. No new large reserves were discovered to justify big investments. By the early 1960s a new dredge cost about £1¼ million and the high prices which would have allowed amortization of such a unit over the limited life of the older alluvial properties could not be relied upon. Meanwhile, gravel pump operations were laid off or resumed according to the state of the market. In the fourteen months to February 1959 the number of
162

Malayan tin mining operations of all kinds was cut from 738 to 386 in order to restrict output to a level permitted by the International Tin Council (I.T.C.) in its efforts to keep up prices. In the mid sixties output picked up, and in the year to May 1966 the work force increased by 6,700, or 16.3 per cent. In the following eighteen months tin prices fell again and 270 operations closed down. Still troubled by their high standing charges, lack of ore reserves and inability to install new capacity quickly, the share of the dredges fell away. However, a number of new dredges have been built in recent years in good locations.

Table 28: Malayan Tin Producing Operations and Methods, 1963, 1965, 1969

	1963	1965	1969
Number of active tin producing operations (at year end)	709	1103	1067
of which			
Gravel pumps	593	979	955
Output (th. tons)	24	31	39
Per cent total	39.7%	48.5%	54.8%
Dredges	66	65	65
Output (th. tons)	27	25	24
Per cent total	45.9%	39.2%	33.1%

Malaya still depends heavily on its two traditional primary products. In 1969 rubber exports were worth £279 million and exports of tin were second at £129 million. The tin duty, an *ad valorem* one tied to the metal's price by a sliding scale, brought in £16.6 million.

Over the last forty years Malaysia and Nigeria have broadly held their old production levels. Thailand has increased its share but Indonesia and Bolivia have fallen away. Costs for most other producers are high in comparison with those of Malaysia. Nigeria has less attractive ore and the mines are located on the Bauchi plateau of the north-centre and so far inland. In the Biafran war they lost most of their labour, had difficulty in obtaining supplies and machinery and in shipping their product. The new Nigerian super tax is said to be so burdensome as to make more capital spending or the introduction of new enterprise unlikely. Nigerian mines have now moved, probably permanently, to a higher level of production costs. Even after the end of the civil war production continued its decline–9.8 million tons of tin in concentrates in 1968, 8.7 million in 1969, 8.0 million in 1970. By 1966 Sukarno's Indonesia was exporting only one quarter as much ore as in prewar days. In part this was because of inadequate machinery, in part due to the confrontation with Malaysia, the trade embargo cutting off Indonesian mines from their Malayan smelters. By 1969–70, under President Suharto, Indonesia was making progress towards a more stable economy and expansion in her tin output was beginning.

The situation of the Bolivian tin industry is especially difficult. The mines of Bolivia are inaccessible. They are concentrated in the eastern cordillera of the Andes at heights ranging up to over 15,000 feet. In the Spanish Empire trade with Europe was largely via the Plate estuary and so involved very long overland connections. Even with the opening of outlets on the Pacific difficulties remained, for the mines are some 250 miles inland. Yet in the cordillera and the altiplano tin mining has populated barren areas or held people in areas which would otherwise have been deserted as Bolivia's old staple of silver decayed.

For the silver of Bolivia the Spanish sacrificed much of the rest of their inheritance from the Incas. Hundreds of thousands of wretched Indians died in the mines, irrigation works were ruined and agriculture declined. The centre of the silver trade was the city of Potosi at the foot of the Cerro de Rico, or Silver Mountain. By 1611 Potosi was said to have a population of 170,000, far in excess of the total for any other American city for many generations to come. In the eighteenth century a surge in Mexican production began to erode the Bolivian leadership in silver and in the nineteenth century her output dwindled as a result of political uncertainties, growing labour scarcity and mine exhaustion. In the mid 1870s one tenth of the Bolivian exports to Britain was still made up of silver but by this time the summit of the Silver Mountain was honeycombed with some 5,000 mines, largely worked out, and the lower workings were frequently swept by flood water. Potosi itself was sadly decayed, its 1880 population being below 23,000.

In 1873 deposits of tin were discovered at Caracoles in La Paz department and the investment of foreign capital and the construction of a skeletal railway system laid the foundations for the prominence of the new mineral. Even so it was not until after 1900 that exports of tin surpassed the value of those of silver. In the early years of the twentieth century Bolivia ranked behind Malaya and the Dutch East Indies, with an output about one eighth the world total. By 1929 her share had increased to about one quarter but was almost exactly one eighth again by 1969.

Even in largely underdeveloped Latin America Bolivia is known as a poor country. Great fortunes have been wrung out of its mineral wealth in this century but its population remains wretched. Until after the Second World War, and excepting the army, Bolivia was, in John Gunther's colourful description, '. . . a kind of "company town" of the tin merchants'.[4] The great mining dynasties, Hochschild, Aramayo, and above all Patino, made their millions and gave Bolivia little in return. Patino invested much of

24. *Catavi Tin Mine, Bolivia.*
Catavi, said to be the world's largest underground tin mine, accounts for about one third of Comibol's output.

165

25. *Colquiri Mine, Bolivia.*
Colquiri is one of the big three tin mines owned by Comibol.

his money in Liverpool smelters and even in low-cost Malayan tin mines, which could undercut the Bolivian ones, though their ore was also necessary to mix with Bolivian concentrates. In 1949 the Patino interests shipped 41 per cent by value of Bolivian tin.

The National Revolutionary Movement (M.N.R.) took over in Bolivia in 1952. It formed the Corporación Minera de Bolivia (Comibol) to take over the three major private groups and conduct three quarters of Bolivian tin production. By the mid fifties Comibol controlled 70 per cent of all the mineral output of the country. Since 1952 it has struggled continuously with a host of problems – environmental, commercial and labour.

Bolivian tin is found in an 800 km. arc running from the Peruvian border east of Lakes Titicaca and Poopo to the boundary with the Argentine. There is a concentration of the biggest mines around the railway junction town of Oruro. The Oruro district produced two thirds of the output in 1899 and in 1964 56 per cent of Bolivia's tin still came from within 100 km. of the town. In the whole tin belt environmental conditions are extraordinarily difficult. As a recent account put it, mining '. . . is carried out often against a spectacular backcloth of alpine peaks and the eternal snows of the glacier-ridden Cordillera Real at heights sometimes over 5,000 metres and in conditions as bleak as can be found anywhere in the world'.[5] In spite of this cold climate, much of the ore is worked at depths where problems of high temperature and humidity are extreme. The almost incredible conditions at the Unificada Mine at Cerro de Potosi were described by P. Schmid in 1957, '. . . the men stand here naked behind their compressed air hammers while water is sprayed on them by one of their colleagues from a hydrant. After five minutes they fling away their tools exhausted and their mates take their places at the steaming stone'.[6] The journey to the coast is over difficult country and the ore then passes through foreign ports. The Chilean ports of Antofagasta and Mejillones are 950 km. from Oruro, Mollendo 1,000 km. away – including a 200 km. journey across Lake Titicaca which involves two extra handlings. Since the mid 1950s Comibol has tried, against considerable odds, to rationalize its mines and reduce dependence on tin. Bolivian ore is of decreasing grade, 3 per cent tin content in 1938, 1.87 per cent in 1950 and 0.82 per cent in 1964. As grades decline so extraction costs go up. Immediately following nationalization, much capital and technical staff fled the country. At the same time the industry was hit by a cost/price squeeze. Heavy new taxes on tin were designed to support the much needed M.N.R. programme of land and social reform. By this time, following the Korean War boom, international tin prices were falling so that by the mid fifties Comibol tin production costs were sometimes about 50 per cent more than the market price.

Labour costs have been an especially difficult problem. The bottled-up grievances of the miners burst out and pushed up costs while new social policies forced Comibol to maintain too big a work-force. In its first two years, while production fell 15 per cent, the payroll went up by one third. In 1952 employment was 25,000 but by 1959 30,000, though at that time Bolivian's tin production quota could have been met with half as many. After reaching a peak of 36,000 workers, Comibol eventually cut the work-force to 28,000 and aimed for 20,000. It began to spend large amounts on new equipment. However, in 1963 there were strikes in the mines and a threat of civil war against this re-organization programme, and in the same year the cabinet resigned in opposition to mine re-organization plans. In 1964 the M.N.R. government fell. Its successor imposed tighter labour discipline, but it was reckoned at this time that Bolivian production costs were 50 per cent more than those of Malaya. There were some extreme cases. By 1965 production costs at the big Catavi Siglo XX mine were 50 per cent above London tin prices.[7] By 1965 Comibol was on the verge of collapse. At this point the miners' leader was arrested, the miners struck and the Bolivian air force attacked their mine headquarters. However, in the mid and late 1960s a more effective labour discipline was maintained. Then in September 1969 a new revolutionary government came to power and labour agitation induced it to try to placate the workers, Comibol being forced to re-employ some labour which it had laid off. In autumn 1970 in a fresh outbreak, left-wing workers seized some of the Oruro area mines.

These recent disturbances and government intervention were expected to turn a Comibol profit of $4.5 million in 1969 into a loss over half as great again in 1970. Poor labour conditions are matched by poor plant. A 1967 World Bank study of Comibol operations showed that some of the concentration mills were in bad mechanical condition and others needed a complete replacement. It noted the imbalance of operations, in some cases modern heavy media separation techniques being followed by wheel-barrow carriage of the product.[8] After 1961, in a so-called Triangular Operation, the U.S., Germany and the World Bank made finance available to overhaul the technical equipment of the Bolivian tin industry, but continuing labour difficulties imperil this overhaul. The continuing flight of capital following the expropriation of more foreign interests and the state monopoly for mineral export, imposed early in 1970, will certainly not help.

Most Bolivian ore is dispatched to Britain as a concentrate of about 50 per cent tin content. As late as 1964, 85 per cent of the tin was exported as concentrates. One way to increase revenue is to process more at home, though, as noted above, the nature of Bolivian ore makes this difficult without an admixture of imported ores. Fuel supplies are difficult. Since the late 1930s and especially since 1947 some ore has been smelted at Oruro, and since 1961 a second smelter has been at work. By 1969 a third and bigger smelter was being built at Vinto. In addition to securing a higher level of processing and therefore more home income generation, Comibol is trying to widen the range of mining. It is involved in special surveys of alluvial deposits of tin, gold and other minerals, plans to double silver production almost immediately, to triple copper output and increase that of bismuth. Further prospects for lead, zinc, copper and wolfram will be assessed. Meanwhile, the poverty-stricken Bolivian economy is still heavily dependent on its struggling tin sector. One third of the country's food must be imported and tin provides two thirds of export income. The value of tin production amounts to about one quarter of the Bolivian gross domestic product. This income is at the mercy of an international tin pricing system which, in spite of the most vigorous efforts for control, results in very wide variations.

International Tin Price Control

The wide variations in tin prices stem from rather distinctive characteristics. In the first place, tin has a very marked price inelasticity of demand stemming from the fact that the major use is in tinplate where tin cost is of so little moment. There has been a relatively slow growth in demand, and much less vertical integration from mine to smelter than with copper or aluminium. On the supply side, it is significant that almost all of the output comes from the primary producing countries with widely differing scales of production and costs. Most mining operations are small as compared with those in other base metals so that decision-taking to increase or decrease output has been highly dispersed. Such conditions tend to widen price oscillations. For almost fifty years attempts have been made to reduce the amplitude of these variations.

In 1920 the year's highest price was £419 a ton but the low was £195. During the six years 1947–53 prices rose from £382 a ton to £1,620 and fell to £566. In 1928–9 a relatively weak Tin Producers Association was set up to try to reduce price variations, but when prices collapsed in the slump it was realized that some more complete coordination was necessary. In 1931 producers of some 90 per cent of the world's tin, including Bolivia, Malaya, Nigeria, Siam and the Dutch East Indies formed the International Tin Control Scheme. This restricted output to support prices, and in 1934 and 1938 operated a buffer stock scheme. The International Tin Control Scheme was abandoned in 1946. In 1950 a new scheme was submitted to a United Nations world tin conference, but no agreement was reached. The need for regulation to give the reliability essential to justify investment in new capacity was, however, realized and in 1956 a five-year International Tin Agreement was drawn up between all producers outside the Soviet bloc and involving all major consumers except the U.S.A. The resulting International Tin Council operated a buffer stock whose manager was given the task of trying to contain prices within a narrow band, buying when prices slipped below a floor price and selling when a ceiling price was reached. In 1957–61 the floor and ceiling were respectively £730 and £880 a ton. At neither end of the scale was the operation successful.

In 1957 the U.S. government stopped tin purchases for its strategic stockpile. At the same time the Soviet Union, having refused to join the Tin Agreement, increased its tin exports to the West, even though in 1958 the British and Dutch governments tried to shield the I.T.C. from the effects by restricting imports of Russian tin. By 1958 world economic activity had taken a downturn and tin prices were falling. At this time Bolivia had to request a U.S. loan of $26 million to help her through the resulting crisis, and in September 1958, after spending £20 million on tin for the buffer stock, the manager's funds ran out and prices fell below the floor price. They reached as low as £640 a ton on the London Metal Exchange. Thereafter, recovery was rapid and in June 1961 the

170

entire tin stock was exhausted and prices rose. As this happened, it was suggested that releases from the U.S. stockpile should also be used to check price rises but this time the president of Comibol, troubled by Bolivia's high production costs, remarked that such a move would constitute economic aggression against the under-developed tin producers.

A second postwar International Tin Agreement covered the years 1961–6 and included six producers and fifteen consumers, though again not the U.S.A. In spite of U.S. stockpile releases, prices rose. The third agreement ran from 1966 to 1971, but in the late sixties the rise in prices in the free market was so strong that I.T.C. again was incapable of checking it for long. In April 1970 its buffer stock ran out yet again followed by a rise to a record £1,630 a ton. Prices fell slightly late in 1970 and the buffer stock manager started to buy again. The I.T.C. was determined to try again. In January 1971 a Fourth International Agreement was concluded. It differs in two major respects from its predecessor. Buffer stock operation is more flexible, and the U.S.S.R. has joined the scheme.

The medium term prospect for tin involves a fall in price to more moderate but still rather high levels. Rio Tinto Zinc has completed a study which suggests £1,200–£1,400 a ton as the price range for the late seventies. Consumption in the non-communist world is unlikely to rise as economy and substitution of other materials goes on, but present high prices are stimulating an important increase in production capacity. Some of this is in the advanced countries, new capacity in Australia or extensions of the small Cornish industry, but there will be big increases in Malaya and Thailand. Indonesian output was halved between the mid 1950s and late 1960s, but with a replacement of out-worn equipment there should be revival there in the seventies.

8. The Political Factor in Mineral Exploitation

PART I

THE U.S.S.R. – MINERAL PRODUCTION ACCORDING
TO A DIFFERENT SET OF RULES

As yet the Soviet Union shows no sign of realizing Mr Khrushchev's dream of outdoing the United States in standards of living, but it is undoubtedly the world's second industrial power. Long continued emphasis on basic industry and the speed of its economic growth have boosted mineral production so greatly that, from a very lowly rank before the Revolution and a middle position among industrial powers at the outbreak of the Second World War, the U.S.S.R. is now, on balance, the world's leading producer of metallic minerals and coal. In oil production it still lags well behind, but projections to the end of the seventies suggest that by that time it may be the world leader there as well.

The U.S.S.R. put great efforts into mineral development in the late twenties and early thirties; the rate of growth of Soviet mineral output between 1926 and 1937 was more than twice that of the fastest expansion period for minerals in the U.S.A. in the twentieth century, that of the years 1902–17. In spite of this, and even though America was still painfully climbing out of the trough of the Great Depression, the value of Soviet mineral production in 1937 has been estimated at no more than 23.5 per cent that of the U.S.A.[1]

In three out of these seven products the U.S.S.R. was well ahead by the end of the sixties, and it led the U.S.A. also in five of the other eight mineral products and metals listed below. Centralized long-range planning, systematic coordinated survey work and the pursuit of mineral self-sufficiency to support a forced industrialization are factors which have transformed the U.S.S.R. into the world leader in mineral production. Yet this achievement has taken place insulated from the forces of the international markets for minerals, and sometimes at high cost when compared with prices ruling there. Shaped though it is by geological facts, and by the
172

Table 29: *Mineral Production U.S.A. and U.S.S.R. 1937 (thousand metric tons)*

	U.S.A.	U.S.S.R.
Coal	448,432	122,579
Oil	172,866	27,821
Iron ore	37,290	14,000 (1936)
Copper	764	93
Lead	422	56
Bauxite	427	250
Primary aluminium	133	45

Source: *Whitaker's Almanac,* 1940.

inheritance from the past, the structure and geographical pattern of contemporary Soviet mining bears the strong stamp of the doctrines of communism.

Table 30: *Mineral Production: World, U.S.A., U.S.S.R., 1969 (thousand metric tons)*

	World	U.S.A.	U.S.S.R.
Coal	2,845,000	507,000	596,000
Oil	2,145,000	457,000	328,000
Iron Ore	701,000	90,000	186,000
(Steel)	(575,000)	(126,000)	(108,000)
Copper	5,870	1,415	900
Lead	3,155	473	440
Bauxite	45,952	1,688	5,000
(Primary Aluminium)	(9,396)	(3,441)	(1,450)
Zinc	5,306	541	610
Tin	199	–	26
Manganese	18	–	7
Phosphate Rock	77	36	15
Potash	16	3	3
(Silver millions of troy oz)	(283)	(40)	(37)
(Gold th. fine oz)	(47,000)	(1,525)	(6,370)
Total for bulk minerals*	5,751,593	1,058,156	1,117,001

* Products bracketed are omitted from the total.

Source: *Mining Annual Review,* 1970.

Marxism and Economic Planning

Marxist theoreticians and modern Soviet writers alike contrast their own scientific methods of resource exploitation with the irrationalisms and contradictions of capitalism. Marx was too concerned with the dynamics of capitalism, which he believed would eventually destroy it, to attempt any full delineation of the economy and society which would follow. However, the Communist Manifesto, first published in 1848, did recognize that the dictatorship of the proletariat would have to make 'despotic' inroads into the old economic and social order in order to 'wrest' power from the bourgeoisie and lay the foundations of the new order. Some of the preconceptions and policies of Soviet planning may be traced back to the Manifesto. Among other points, it listed state control of all means of transport, extension of factory industry, the 'bringing into cultivation of waste lands', and the 'combination of agriculture with manufacturing industries, gradual abolition of the distinction between town and country by a more equable distribution of the population over the country'. These constitute the root of the Soviet concern to spread development over the national territory and of the belief that in a scientifically conducted state, man could master nature. Engels emphasized that planning was essential in order to attain a rational pattern of economic activity. As he put it in *Anti-Duhring*, 'Only a society which harmoniously combines its productive forces in accordance with a single overall plan can permit industries to be distributed throughout the country in a way most favourable to their own development and to the preservation and development of the remaining elements of production.' The overall framework which Engels envisaged certainly shaped Soviet mineral development. Competition between districts was to cease, development was to be rationalized and to be spread more widely. In the late 1950s two leading Soviet geographers, P. M. Alampiev and Y. Saushkin, summed up the contrast with the West and some of the implications. Alampiev suggested that in the West, agricultural, mineral and industrial areas were formed '. . . spontaneously, in the process of the
174

anarchic development of the economy. Under the conditions of a Socialist society, the economy is conducted according to plan.' Saushkin asserted, 'one of the most important principles of developing the national economy of the Soviet Union is an all-round comprehensive development of national resources'. The location, scale and marketing policy of mining was affected by this economic philosophy. Minerals are worked in locations which even a modern, dynamic, capitalist Russian state would not have chosen.

Neither Marx and Engels nor, indeed, Lenin laid down very definite guidelines for a socialist state. However, soon after the Revolution the rudiments of an overall framework for development began to take shape. Even in the period of war communism (1918 to spring 1921) a long-term economic programme and some of the basic principles and machinery for planning were hammered out as the leaders made the difficult transition from securing the overthrow of one state organization to the operation of a new system for its successor. The ideological base of fear of, and implacable opposition to, capitalism was now reinforced as a result of direct intervention by western armies. This experience was not forgotten in the period of the New Economic Policy, and in 1931, in the middle of the first Five Year Plan, Stalin observed, 'We are fifty to a hundred years behind the advanced countries. We must make good the lag in ten years. Either we do it or they crush us.' The emphasis on heavy industry which this attitude encouraged made rapid development of the mineral sector imperative. Coal and iron ore grew most spectacularly, but non-ferrous metal production had to increase rapidly as well. The mineral industry inherited from Imperial Russia provided an inadequate and spatially extremely ill-balanced base for this development.

Writing in 1899 in *The Development of Capitalism in Russia*, Lenin pointed out that the substantial growth of capitalist industry over the previous quarter century had opened up much of the south and the east of the empire as an area which was effectively in the relationship of a colony to the centre and St Petersburg. Agricultural and mineral raw materials from these outlying regions fed the factories and populations of the industrializing core of

Mineral Resources

European Russia. In 1897 the total value of Russian industrial output from all except small handicraft or domestic industry was 2.8 thousand million roubles. A survey made in the previous year showed the following areal breakdown: Moscow region (six provinces) R755 million; St Petersburg Province R317 million; Poland (three chief provinces) R335 million; Ukraine (four provinces) R246 million; Urals R85 million; Baku area R82 million; South-west European Russia (two provinces) R135 million.[2] By 1913 the degree of concentration had if anything increased. By that time about half of the industrial output came from Central European Russia and the St Petersburg area, 18 to 20 per cent from the Ukraine/Donbass. The whole eastern three quarters of the territory, including the Urals, produced only 11.5 per cent of the total.[3]

In one or two cases only did these pre-revolutionary centres of high economic development have a local rich mineral resource base – in marked contrast to the coal-based industrial conurbations of Western Europe. The Ukraine or Southern Industrial Region was an exception, built up essentially after the 1870s on the basis of Donbass coal and Krivoi Rog iron ore. The inadequacy of mineral resources in the key industrial regions of Moscow and St Petersburg and the absence of industry in the outlying mineral districts was a reflection of peculiar Russian circumstances and was, as Lyaschenko put it, 'extremely irrational in character on the whole'. In part it reflected ignorance of the resources of the empire, but ignorance in turn was related to the peculiar business climate. Other factors according to Lyaschenko were the perpetuation of the old semi-feudal pattern of industry derived from the eighteenth century, the Russian policy of discriminating against non-Russian areas and the role of big concerns, 'monopoly capital', which opposed the opening of new and competing districts. An example of the latter was the effort of Baku oil firms to prevent development in competing areas further north. On the other hand it was already becoming clear that the empire was rich in minerals.

The Geology and Mineral Development of Imperial Russia

The broad pattern of geological structure in the U.S.S.R. suggests a basically highly favourable mineral endowment. There are two core areas of basement rocks; the shield of Baltica, shared with Scandinavia, and that of Eastern Siberia. Both underlie a much wider area than is indicated by their surface outcrop. These shields are rich in vein minerals and in bigger occurrences of metallic minerals, both ferrous and non-ferrous. On the fringes of these geologically very old and rigid platforms, and, in the case of Eastern Siberia in various parts within it as well, are basins of thick sedimentary deposits containing oil, coal and natural gas. The Urals separate the two shield-cored areas, and much more impressive mountain ranges fringe the country on its long southern and Pacific boundaries. Igneous masses have been intruded into these fold mountains. The effects of the resulting major mineralization process are exposed where denudation has been severe.

This mineral wealth is contained in a country which has approximately one sixth of the earth's land area outside the ice caps. The U.S.S.R. is roughly two and a half times as big as the U.S.A., its extreme dimensions from the Carpathians to Kamchatka being as far as from San Francisco to London. Most of it has a harsh climate, much is covered with poor soil, coniferous forest, or swamp land, so that tracts as large as Western Europe have scarcely any permanent population. In the these circumstances and even discounting the slowness of economic development and the aberrations of 'monopoly capital' it is hardly surprising that mineral resources were slow to be recognized and developed.

The average coal output of Imperial Russia in 1901–3 was half that of France and just over 7 per cent that of the United Kingdom, from which it annually imported two and a half million tons. The 1905 copper output was a mere 8,700 tons or 2.1 per cent of the American level. At that time 70 per cent of Russian copper consumption was supplied by imports, though over the next few years there was growth in home production. Between 1901 and 1903 the empire was not among the world's leading fifteen lead producers, its zinc output was insignificant, and no tin was mined. It paid R8.5 million in 1911 for imported tin. By 1913 Russia ranked sixth in world coal production, fifth in iron ore and seventh in copper.

Much of even the relatively small home production and the import trade in minerals was in the hands of foreign capitalists. London merchants dominated the trade in tin into the metal districts of southern Russia while, just before the First World War, Berlin and Hamburg capitalists were planning to open up tin deposits known since 1811 in the Trans-Baikal region.[4] At this time too German engineers were investigating silver/lead ores in the Onega district, and in the previous summer four expeditions had investigated various western Siberian mining properties for French financiers. Russian minerals were worked by some 135 British companies with investments totalling £28 million, £17 million of this in Caucasian oil, £7 million in gold and £2 million in copper. Foreign exploitation was not only well established, but also blatant. The 1913 *Russian Yearbook* paraphrased a Belgian report made at the end of a two-month survey of the wealth of Siberia: 'Mining companies registered in accordance with Russian law, and having a Russian director, may in practice be composed of foreigners, and foreign material may be used without violating the letter of the law.'[5]

At this time production was heavily concentrated in the west. It was already known that Siberia was rich in minerals though knowledge was vague. A wide scatter of gold workings was the most characteristic feature of the mining map of the area east of the Urals. There were the merest beginnings of coal working in the Kusznetsk Basin (Kuzbass) and in the Karaganda area of Kazakhstan. In Kazakhstan too were the Uspensk copper mines – an important British investment – and the lead/zinc operations at Ridder (now Leninogorsk). There had been iron works in the Urals since the early eighteenth century, but the days when they dominated production were now long past, techniques were extraordinarily antiquated, and their operations were so uncompetitive with the new coal-based works of the Ukraine that their share of Russian iron production fell from 70 per cent in 1860 to only 20 per cent in 1913. Copper production remained centred in the Urals, but the

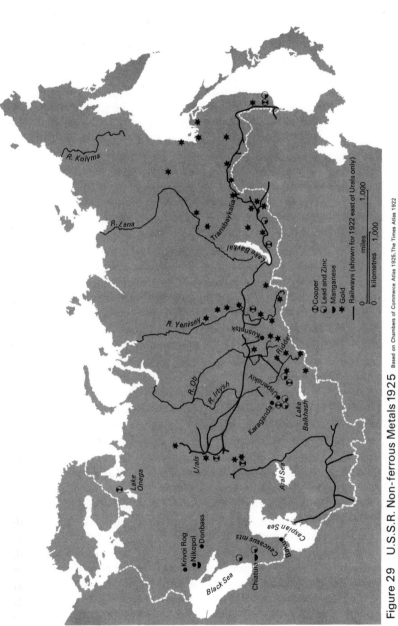

Figure 29   U.S.S.R. Non-ferrous Metals 1925   Based on Chambers of Commerce Atlas 1925; The Times Atlas 1922

mines were inefficient and costs increased as Russia fell from its old position of a world leader in copper to the status of a minor producer.

Soviet Mineral Development

The primary Soviet task was to increase the power of the state from its own resources. Another aim was to spread economic development across the map. Important as a development philosophy, rather than an aim, was the conviction that socialist man could make any development successful, even if to bourgeois economists it was not viable. Coal and iron received the greatest attention as a result of the emphasis on basic, strategic heavy industrial products, and above all because of an understandable if rather megalomaniac preoccupation with steel expansion. Looking back in 1933 to the recently completed first Five Year Plan, P. Savitsky summed up the geographical implications of Soviet policy with great simplicity. Two basic elements which he identified were '. . . the question of emancipating the central regions from imported foreign raw materials and fuel by a rational utilisation of their second rate natural resources; and that of establishing a new industry based on the first class resources of the outlying regions'.[6]

The essential first element in the Soviet mineral programme was large-scale, integrated geological survey. Expenditure on geological survey rose from R187,000 in 1906 to R10 million in 1927–8 and R141 million in 1932. In the mid 1950s Academician Obruchev was said to recall the time when he was the only geologist working between the Urals and the Pacific, and at the Revolution only 0.45 per cent of the national territory had been surveyed at a scale of 1:100,000 or larger. By 1938 4.5 per cent of the U.S.S.R. was geologically surveyed on a 1:100,000 scale or larger, a situation incredible in Britain three quarters of a century earlier. These surveys were not haphazard ones like those of previous periods but 'complex organizations which methodically realize a single plan of investigation of the country'. Such a rapid expansion of geological work brought problems, including use of inferior equipment and dilution of skills, but the results were spectacular. By the mid 1930s it could be claimed, and fairly so in spite of the frequency of ideological hyperbole, that '. . . a revolution has taken place in the former estimates of the mineral resources of the country'.[7] A bald recital of major discoveries is impressive enough.

East of the Urals major coal and iron ore fields were either discovered or more fully delineated. In the twenties the geochemist Fersman proved the existence of huge apatite deposits in the Khibin Mountains of the Kola peninsula, deposits rich in phosphates and in nephelite, a potential source of aluminium. Potassium salts were unknown before the Revolution, but in 1925 deposits were proved at a depth of only 92 metres over 40 square kilometres near Solikamsk in the north-west Urals. Poland had been the sole source of sulphur in the days of the Russian empire, but after 1929 big discoveries were made in Central Asia and along the middle Volga. Copper resources were supplemented in 1928 by major discoveries at Kounrad near Lake Balkhash. Lead, almost wholly from Caucasia before 1917, was proved in workable deposits in Siberia and the Far East, and in 1930–31 copper, lead and zinc deposits were found in Ferghana, Central Asia. The very important chrome-nickel-iron deposits of Khalilovo at the southern end of the Urals Economic Region were proved in 1928. Although the late twenties and early thirties was the period of marvels in Soviet mineral development, important finds continued to be made throughout the thirties and in the postwar years too – as with the great iron ore resources of Kustanay on the Tobol or, in the mid fifties, the discovery in Kazakhstan of iron ore reserves said to be on the scale of the Lake Superior deposits.

Whereas before 1917 the mineral products of the east were mostly derived from a scatter of small gold camps, now the map of mining both widened and filled out. The aluminium industry, whose progress may be compared with that in the west, illustrates some of the problems and policies in Soviet mineral and metal development. Finally, a range of current difficulties will be considered.

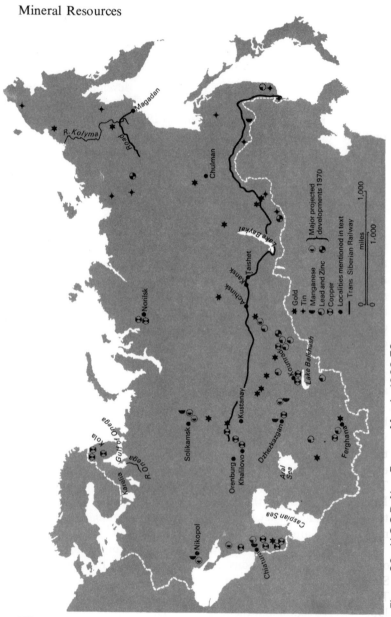

Figure 30 U.S.S.R. Non-Ferrous Metals 1969-70   after T. Shabad 1970

The Development of the Soviet Aluminium Industry

Imperial Russia paid no part in the early development of the world's aluminium industry. No resources of bauxite were known. The U.S.S.R. made its first aluminium in 1932. By 1941 its capacity of 100,000 tons ranked it fourth behind the U.S.A., Germany and Canada. North American capacity expanded rapidly in and after the war, and by 1953 Shimkin observed that the whole Soviet sphere, with about 180,000 tons capacity, was equal to only 16 per cent or so of North American output in 1944, and came to the happy conclusion that this rendered the U.S.S.R. weak in 'air power potential'. It seemed a back number in aluminium '. . . hampered by weak domestic bauxite resources, a late start in aluminium production, poor aluminium technology, inadequate available power, and war-time destruction of part of its aluminium industry'.[8] Much of this adverse comment was justified, but, as always, Soviet planners were working along different lines of economic rationality from the west and they tackled the problems of aluminium with vigour. By 1970 the output of the U.S.S.R. was 1.5 million tons, well in advance of all producers except the U.S.A. with its 3.6 million ton production. The rest of the communist bloc added another 0.5 million tons.

The Soviet Union is poorly endowed with good quality bauxite, but, like Nazi Germany, realized that an aluminium industry was vital for military purposes with air power coming into its own. In the thirties the Germans supplemented their own poor resources by exploiting the Balkan bauxites and, above all, those of Hungary. The Russians began with the bauxite deposits of north European Russia, but found that these were both of poor quality and exposed to military attack. They moved on to eastern deposits, to substitute minerals and, after the war, to use Balkan bauxite.

Completion of the Volkhov power station, east of Leningrad, was one of the first major industrial achievements of the new Soviet State in the mid twenties. Fifty miles east of Volkhov, in the Tikhvin area, were the only bauxite deposits known until after 1930. The reduction unit then built at Volkhov survives, but is now the smallest of seventeen aluminium plants. This operation had two major weaknesses. The high silica content of Tikhvin bauxite involved high production costs – in 1938 world average practice required five to six tons of bauxite for every ton of metal, but it is said that only 44,000 tons of metal were produced from 560,000 tons Tikhvin bauxite.[9] The second disadvantage concerns real power costs. In the Leningrad industrial area of over three million people, power availability is small in relation to demand so that such a power-intensive industry is not a suitable specialism.

The second Russian unit was opened in 1933 at Zaporozhe near to the much bigger Dneprovsk power site on the Dnepr. Even so the plant was on the edge of the major Ukrainian market and has had to compete with other major even if less power-intensive industries such as electrical steelmaking. Zaporozhe was dependent on very long hauls of either bauxite from Tikhvin or alumina from Volkhov. In 1940 it produced 36,000 of the country's 60,000 tons of aluminium. The Germans rapidly overran this area, and the Zaporozhe reduction plant was not reopened until 1955. It then began to use Greek bauxite carried to the works by sea. Current production is of the order of 100,000 tons primary aluminium annually.

The location assessment for new aluminium reduction plants began to change in the thirties as further mineral resources were found. By 1931 it had been realized that alumina could be produced from the nephelite in Khibin apatite, and in 1933–4 new bauxite deposits were discovered in the Urals and in the Angara area of central Siberia. The larger bauxite deposits of the eastern Urals north of Sverdlovsk, surveyed in 1938, proved both richer and more uniform in composition than those of Tikhvin. Urals power potential was only mediocre but, on the other hand, the Urals Economic Region was rapidly developing into the second steel base of the country and becoming an important metal fabricating area. In 1939 an aluminium smelter was completed at Kamensk, south-east of Sverdlovsk. Two years later the German invasion caused removal of equipment from the Volkhov and Zaporozhe plants to a new smelter at Krasnoturinsk, and a start on the construction of the first plant further east at Novokuznetsk in the Kuzbass. The Urals now became the chief centre of bauxite,

184

alumina and aluminium production. The Severouralsk area north of Krasnoturinsk has remained the leading centre of bauxite production and, as with Tikhvin ore in the thirties, bauxite or alumina from the Urals is now carried widely to reduction plants in other areas. The Novokuznetsk plant, for instance, completed only in 1956, used Urals alumina railed over a distance of 1,300 miles until 1964 when supplies became available from Pavlodar only 450 miles away in eastern Kazakhstan. In 1942 the Urals smelters produced more primary aluminium than the country's prewar total. By 1968, however, Kamensk and Krasnoturinsk had only one fifth of the country's capacity. As in North America, postwar expansion in aluminium has involved a search for major new power sites. In the U.S.S.R., additionally, the availability of raw materials has had important locational influence.

Use of alumina, derived from Khibin nephelite, began on a small scale in the late 1930s. In 1949 the Volkhov alumina plant turned over to nephelite, though it also began to use Tikhvin ore again in 1953. In 1939 work was begun on an aluminium plant at Kandalaksha at the north-western extremity of the White Sea. Interrupted by war, construction was resumed in 1949 and the Kandalaksha works was completed in the early fifties. Another north-western smelter was built at the same time at Nadvoitsa on the Baltic White Sea Canal between the Gulf and Lake of Onega.

At the end of the thirties it was realized that mineral deposits and power availability alike made the Angara River area and Azerbaidzhan S.S.R. attractive smelter locations. In 1950 the industry built a works at Kanaker to use power from hydroplant on Lake Sevan, and in 1955 at Sumgait, north of Baku, employing Caucasian power. Both used Urals alumina, but a much nearer source of supply became available in 1966 when a plant at Kirovabad began to produce alumina from alunite.

Production beyond the Urals began with the Novokuznetsk plant in the Kuzbass coalfield, but the development of very large hydro operations on the Siberian rivers was accompanied by still bigger smelters, as at Shelekhovo, near Irkutsk (1962), Krasnoyarsk (1964) and Bratsk (1966). By 1968 these three plants had 36 per cent of estimated U.S.S.R. aluminium capacity. All the Siberian plants –

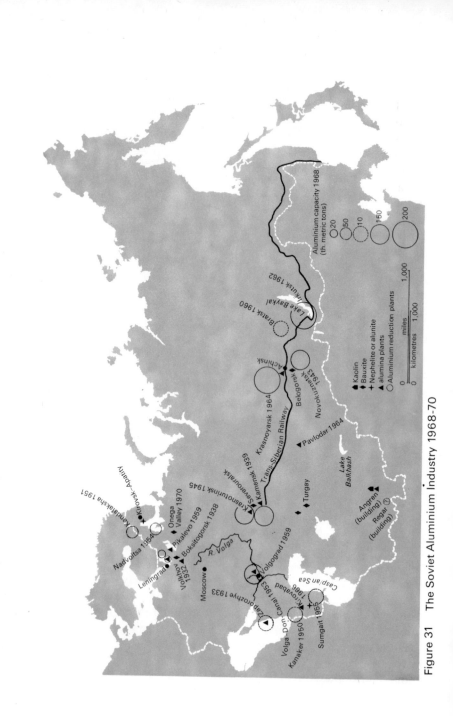

Figure 31   The Soviet Aluminium Industry 1968-70

that is including Novokuznetsk – produced 35 per cent of the country's aluminium in 1965 but were soon expected to have 65 per cent of a much increased total. Although their power costs are low, these works were ill-placed for alumina supply and, even more than the plants of the Pacific north-west of the U.S.A. and like Kitimat, are far from the chief markets for aluminium. Until the mid sixties all their alumina was from the Urals. After this, bauxite deposits were opened in northern Kazakhstan and alumina was being made much nearer to the reduction plants at Pavlodar. Late in 1969 a nephelite to alumina plant at Achinsk went into production. Achinsk has extremely low cost coal supplies, is excellently placed to supply the Krasnoyarsk smelter, and provides further transport economies in alumina supplies to Bratsk and Shelekhovo as well. Finally, bauxite deposits which will also supply the big central Siberian smelters have been discovered at Ibdzhibdak. In accordance with long-established Soviet principles of spreading development throughout the national territory, even if this involves use of poorer materials and higher costs, a reduction plant is now being built at Regar in the Central Asian republic of Tadzhikistan to work on alumina derived from kaolin.

Traditionally the U.S.S.R. has striven to be free of dependence on outside mineral supplies, but the poorness of its resources for aluminium production has caused it to modify this policy. Even though their own bauxite was better than that of Tikhvin, Urals alumina plants in the 1950s were annually processing as much as half a million tons of Hungarian bauxite railed over approximately 2,000 miles. In 1955 this movement was recognized as uneconomic and discontinued. The opening of the Volga–Don canal in 1952 and the completion of the 2,500 m.w. hydro-electric plant at Volgograd (formerly Stalingrad) in 1958–61 created a new and more favourable situation. A Volgograd smelter began operations in 1959 and by an agreement of 1966 alumina from Hungarian plants is carried to it via the Danube, Black Sea and the Volga–Don canal. In return the U.S.S.R. supplies aluminium to Hungary.

As the quantity, quality and location of its own resources have proved inadequate, Soviet alumina imports have grown. It is estimated that the U.S.S.R. produced 5.2 million tons of bauxite in 1969, one million tons of nephelite concentrate and one million tons of alunite ore. It imported about 1.8 million tons of high grade bauxite and alumina from Hungary, Yugoslavia, Greece and even the U.S.A. In the course of 1970 Soviet interest was expressed in bauxite development in West Africa. As the economic advantages of international sources of supply become clearer the old economic self-sufficiency of the U.S.S.R. is weakening. For other minerals there is little sign of this trend.

Meanwhile, efforts are being made to improve the aluminium materials supply situation at home, but progress is slow and a number of weaknesses in Soviet planning are showing. For instance, at a time when the high cost of Siberian dependence on bauxite and alumina from the Urals was becoming obvious, the first part of the new open-pit bauxite operations at Turgay in north-western Kazakhstan was put into operation. Since then, development work has slowed down, capital equipment there being unused or under-used year after year. Production in the Turgay pits in 1969 was only 60 per cent of the target figure. As a result the new Pavlodar alumina plant, itself behind schedule, receives inadequate and poor supplies of bauxite. The same planning difficulties apply with the smelters. Bratsk began work in 1966, but is reported to have operated at only one third of its planned capacity until the end of 1969. In other words, in real cost terms its investments are not being covered. In the west discounted cash flow analysis would show how extremely unsatisfactory such a situation is. It points to grave weaknesses in the Soviet planning system. All the undoubted success in terms of mineral and metal tonnage increases has been dearly bought.

Soviet Planning Problems in Mineral Development

Some of the basic problems of Soviet mining are derived from the Marxist–Leninist approach to the physical environment. Others stem from the structure of economic enterprises and yet others from difficulties in marshalling the necessary factors of production, above all men and capital.

Marxism asserts the prime importance of the structure of society and economy in determining the occupation and the wealth of nations and areas. It has, therefore, always rejected environmental determinism. In doing so, it has been in danger of neglecting the importance of physical impediments to economic development. It is true that Lenin realized clearly enough that in so huge a state it was impracticable to adopt uniformity in development policies. As he put it to the tenth Party Congress in 1921: 'To standardize Central Russia, the Ukraine and Siberia, and to subject them to a set pattern would be the height of folly.' But his followers found it tempting to ignore this warning, especially as there was a prevalent conviction that socialist man could achieve any object. The development programme for the north and later the Virgin Lands campaign in Kazakhstan were products of this philosophy. Mining takes place in very various environmental conditions and it is evident that, by western standards, physical difficulties have not been taken fully into account in all cases.

The Kolyma district of north-eastern Siberia was well beyond the limit of the scattered gold workings of the Russian Empire. Gold was discovered in 1931 and later tin mining was begun in the same area. The construction of a road link to Magadan on the Pacific coast eased conditions of access, but the adversity of climate and of surface conditions could not similarly be overcome. Costs were reduced by the use of forced labour, working under appalling conditions and controlled by the secret police. This slave labour force was applied first to surface operations and then to mining. Forced labour in the U.S.S.R. was abandoned in 1955–6 and immediately labour costs rose. Workers drifted away and the turn-over rate was high so that in 1965 the Magadan Deputy to the Supreme Soviet was complaining of '. . . the flight of cadres and in the first place of mining specialists, geologists, agriculturalists and also skilled workers'.[10] Gold mining adjusted. As wages rose so new machinery was introduced to enable the removal of a thicker overburden so that open-cast operations could be revived. For tin, abolition of forced labour left less room for manoeuvre. All of the seventeen lode mines and thirteen of the placer mining operations were closed down, only the Omsukchan tin mine surviving in the area.[11]

189

Norilsk, east of the mouth of the Yenisei River and at latitude 70°N., is an example of the successful long-term growth of a major mining operation in an extreme environment. Rich copper–nickel ores were discovered in the mid 1930s. Costs of extraction and of shipping proved high, but even though other much more accessible nickel and copper ores were discovered, decline of Norilsk was averted because of the rich content of ancillary metals in its ores – tellurium, titanium, platinum, and so on. The population of the neighbourhood was already almost 14,000 when the first nickel matte was produced in 1939. By the late 1960s, at 150,000, it was far and away the world's biggest agglomeration in such a northerly latitude.

Success under such conditions has been dearly bought. There is a rail link to the Yenisei river. Some 70 per cent of Norilsk production is still from open pits, the winter is characterized by a long period of darkness and a January average temperature of the order of minus 40°F. To attract and to compensate labour for the disadvantages of work in such an area, wages are high. Pay scales in Central Siberia are said to be generally 15 to 20 per cent above the level of European Russia and those of the northern district 50 to 100 per cent above. Under conditions of extreme cold, breakage of metal in equipment is more common. Even so much of that used in the far north is not specially adapted to these extreme demands. An extraordinary picture of conditions was described early in 1970 by V. Loginov: 'In the placer mine of the north, a bulldozer usually requires a complete overhaul after five to six months. Such an overhaul in the north costs about twice as much as a new bulldozer, including delivery charges.'[12] The contrast with the current extremely detailed analysis of various means of cost reduction in major open pits in the western world should be noted (see p. 44). Loginov recognized it frankly enough.

In view of the competition under capitalism, machine manufacturers are ready to fill orders for special-purpose equipment according to customer specifications, even if only small lots are involved. Mining companies and construction and transportation concerns make extensive use of non-standard equipment that has been manufactured according to individual orders for specific operating conditions. The cost of such non-serial equipment is, of course, much greater than that of standard equipment, but its increased productivity soon compensates for the higher purchase cost.

In the U.S.S.R., on the other hand, although there is some equipment made 'to strictly individual designs' it 'is not being mass-produced for specific regional conditions'.[13] Production of specialized machinery, though highly desirable, might need closer liaison between industries than is easy under Soviet planning.

Another problem, and a marked difference from conditions in the West, concerns pricing. Most mineral prices in the capitalist world reflect the play of supply and demand situations around the basic costs of a particular mineral industry – that is overall outlays plus a profit level which pays due regard to the riskiness of mining. High cost producers are drawn over the line and into production when prices are high and at this point the better deposits yield a greater profit. There are managed mineral prices, but as the operations of the International Tin Council have shown, even very careful attention cannot control very wide variations in price.

In a socialist society, land, capital, profit and losses are alike the concerns of the state, and the allocation of a money figure to inputs and outputs is by no means so simple. Soviet planning attempts to ensure that production and consumption are neatly dovetailed, but in these circumstances, what should the price be, and what determines which deposits should be developed or expanded? Although the problem affects the whole economy, it is especially serious in mining, where quality and location of the ore body and conditions of work vary much more widely than production conditions in manufacturing.

It is argued in the U.S.S.R. that differences in costs derived from physical differences should be 'neutralized'. Commonly this involves a continuing redistribution of funds from profitable to unprofitable operations in order to maintain the flow of production. A large number of mines work to planned deficits, being reimbursed by cross subsidization. Some Soviet mining economists have suggested that mineral prices should be based on costs at the worst operations, with the lower cost producers making payments to the state. However, the range of costs is so great that if this principle was applied prices of minerals to manufacturing trades would be

191

excessively high. In 1960 for instance, four fifths of mines producing zinc had deficits, the range of profitability being from a 51.8 per cent deficit to a 100 per cent surplus on the operations.

Some of the intellectual turmoil involved in breaking out both from the crust of Marxist–Leninist dogma and from the practical strait jacket into which it has forced Soviet planners, may be sensed in the 1968 words of a member of the Leningrad Mining Institute. 'If social needs are adequately met by the production of deposits that are average or better (in terms of mining conditions and location), there is no economic need for working the poorer deposits.'[14]

The speed of Soviet industrialization, particularly since it has so conspicuously matured in the 1960s, and the widening of economic effort away from the traditional basic trades have caused shortages of capital and supplies as well as of labour. In the mining sector the great width of the development front has exaggerated these problems. The situation at the end of the sixties was summed up by a western commentator, basing his conclusions on Soviet reviews.

The U.S.S.R. continued to experience considerable difficulty in completing mineral projects on schedule as efforts were dispersed over a large number of projects. Construction sites were inadequately supplied with materials, particularly rolled metal and piping; funds were not sufficient for the volume of installation work; labour-consuming work was insufficiently co-ordinated . . . Many plants and mines were put into operation in spite of numerous imperfections and insufficient equipment. As a result labour productivity was below planned levels, and large numbers of personnel were occupied in repair and auxiliary operations. Many enterprises and projects have operated over long periods with lower capacities than originally planned.[15]

There are many examples of delay in completing projects or in the rounding out of unbalanced operations.

In 1940 the Dzhezkazgan copper deposits in Kazakhstan were first opened on a commercially significant scale. The ore was shipped to smelters in the Urals and on Lake Balkhash. After 1954 the ore was concentrated before shipment. By the mid sixties Dzhezkazgan was yielding about half the country's copper, but it was not until the end of the decade that a long-planned smelter was built there and completion of this has been delayed. Important bauxite deposits were discovered in the Onega valley of northern European Russia in 1949. To the west, at Volkhov and in Karelia, were three important aluminium reduction plants suffering from a shortage of good ores, but opening of the Onega bauxites was not undertaken until 1970. Remoteness is clearly not the explanation of this delay, for the Vologda to Archangel railway line has been in operation a little way to the east of the Onega since before the Revolution and a track now exists around the Gulf of Onega and through to Kola. Development of the low grade but large coal deposits outcropping along the Trans-Siberian railway in the Kansk–Achinsk basin east of Krasnoyarsk has been held back by shortages of both capital and men. A final example is the case of the nephelite-to-alumina installation at Achinsk. Work began on this in 1956, but it was not brought into production until late in 1969. At the end of the 1960s shortages of capital, labour and equipment caused the U.S.S.R. to turn to the possibility of joint development of some of its natural wealth with raw material deficient capitalist economies.

Joint Soviet-Capitalist Development of U.S.S.R. Minerals

Early developments involved negotiations with Japan. The Japanese G.N.P. has been going up ten per cent annually and her industrialization has been so rapid that, as she is deficient in most mineral resources, she has become '. . . the most dynamic and powerful force in world mining today'.[16] Japanese copper consumption in 1965 was 0.44 million tons, by 1968 0.75 million and projections to 1975 suggest 1.21 million. Already by 1969 Japan was the world's leading importer of iron and nickel ores and of copper and zinc concentrates. In their consideration of world sources of supply Japanese firms naturally looked across the seas of Japan and Okhotsk to the only partially developed mineral riches of eastern Siberia. Japanese and Russian needs seemed to dovetail. In high-level talks in Moscow and Tokyo the Russians requested help with major projects – for instance, in the completion of the steel works at Taishet – and supplies of materials and skills. In return the Japanese could obtain long-term mineral supplies. By 1969 the U.S.S.R. had proposed the joint development of the big copper deposits at Udokan north-east of Lake Baikal, and had asked for assistance with natural gas, potassium and phosphates and in port development on the Pacific. Big coking coal developments near Chulman in the southern part of the Yakut A.S.S.R. have been another major theme for discussions. Undoubtedly, some of these joint ventures will go ahead, but the progress of negotiations has not been smooth.

The U.S.S.R. has asked for extremely protracted credit terms – so protracted in fact that the Japanese find it difficult to entertain them. Soviet negotiators have made it clear that their interests will dominate in the arrangements. Moreover, the U.S.S.R. has already shown a readiness to break off mineral supplies when home demand is insistent. In 1967 and 1968 Japanese imports of zinc ore from the U.S.S.R. averaged 20,000 tons. Anticipating a long-term growth of this trade, some of the leading Japanese zinc producers proposed to build new smelters along Japan's underdeveloped west coast for easier contact with the shipping ports. In the summer of 1969 the U.S.S.R. suddenly cut off the flow.

By 1969–70 western nations had become involved in a series of discussions about joint ventures. At the end of September 1970 the C.B.I. had extensive talks with a Russian delegation. Some of the proposed investments were in manufacturing plants, but the British Steel Corporation is apparently interested in iron ore prospects in northern Russia and Rio Tinto Zinc is said to be planning to arrange large investment (possibly in excess of £400 million) in nickel ore development in the Orenburg area of the southern Urals and in Udokan copper, the scheme for which Japanese help was sought only a year before. In October 1970 the visit of President Pompidou to Moscow was also accompanied by talk of joint resources development. As the communiqué at the end of the visit put it, there was a decision to reach agreement on long-term supply contracts '. . . especially in connection with the exploitation of new mineral resources which would lead to a privileged cooperation between the two countries'.

Whatever its past costs, it cannot be denied that Soviet policy for mineral development has been spectacularly successful in terms of output figures. The current topping up with elements of capitalist participation suggests its economics are more suspect. This would have disturbed Marx and Lenin, and would have been incomprehensible to Stalin.

9. The Political Factor in Mineral Exploitation

PART II

THE NEW BRITISH ALUMINIUM INDUSTRY—A
CASE OF CHANGING THE RULES IN THE COURSE
OF THE GAME

In 1969 the United Kingdom produced under 34,000 tons of primary aluminium as compared with approximately 1,450,000 tons in the U.S.S.R. However, by autumn 1971 Britain had plant capable of some 300,000 tons annually and by 1973 its capacity will be 350,000 tons. The form of intervention has been quite different, the capitalist, international aluminium companies retain their independent existence, but Britain's impressive new industry, like that of Soviet Russia, is the outcome of political decisions.

By the late 1960s, Britain had developed a considerable aluminium fabricating business, increasingly dependent on imported aluminium. Its only primary producer, British Aluminium, had turned progressively to more reduction overseas. In its initial development in the second half of the 1890s British Aluminium opened up bauxite in Antrim and also prepared alumina in Northern Ireland. However, '. . . a few years of operation demonstrated that nothing but fixed charges remained from a hasty investment in recalcitrant Irish ore'.[1] B.A.C. then acquired French bauxites. It built an electrode manufacturing plant at Greenock, its reduction plant was at Foyers and a fabricating unit on the site of the old Milton reduction works in Staffordshire. In 1905 a second aluminium plant was constructed at a better location for unloading alumina – Kinlochleven. Another alumina plant was built during the First World War, this time at Burntisland on the Firth of Forth, and at the end of the twenties B.A.C. put up its third reduction plant at Lochaber on the outskirts of Fort William. An aluminium rolling mill built at Greenock was not successful so that all B.A.C.'s finished rolled aluminium came either from Milton or a works at Warrington.[2] Meanwhile in 1909 an independent concern, the Aluminium Corporation, installed both a reduction plant and a

196

rolling mill at Dolgarrog in the lower Conway Valley – but with its alumina plant at Hebburn, Tyneside. Dolgarrog proved to have much lower survival value than B.A.C.'s highland works, and all production there has long ceased. The only two other reduction units ever operated in twentieth-century Britain were practicable only under wartime conditions and were at Resolven in the Vale of Neath and at Port Tennant near Swansea.

As demand grew it was soon impossible to meet it from British reduction plants, and B.A.C. turned overseas for its primary metal as it had done earlier for bauxite. By 1908 it was building Norway's first smelter in Stang Fjord, eighty miles north of Bergen. Four years later another Norwegian plant was purchased. Even with completion of Lochaber in 1930, British capacity could not be expanded significantly beyond 30,000 tons. Imports grew, and as foreign production became more and more competitive, Foyers was switched to quality products, particularly super quality aluminium in which higher costs were rather less significant. However, in 1967 the Foyers plant was closed.

26. *Ardal aluminium smelter, at the head of the Sogne Fjord system.*

After the Second World War much of Britain's imported primary aluminium came from Canada, and in the mid fifties British Aluminium was one of the two partners in the construction of the Baie Comeau plant on the lower St Lawrence estuary. Purchase of Canadian ingots involved dollar currency so that alternatives were sought. The early British interest in the Volta River Scheme[3] proved abortive, and Britain became increasingly dependent on Norway, though Canada led in tonnage until the first half of 1968. Norway then became Britain's chief supplier.

Norwegian 1940 capacity was 50,000 tons, only 50 per cent more that that of the United Kingdom. By 1970 it was 535,000 tons and will rise to 700,000 by 1973, with the construction of possibly three more plants. Important though the British outlet is, the E.E.C. is now an even bigger purchaser of Norwegian metal.

Table 31: *United Kingdom supplies of primary aluminium, 1950, 1960, 1967* (*thousand metric tons*)

	1950	1960	1967
United Kingdom production	27	28	39
Imports of ingots etc.	138	306	303
supplied from			
Canada	118	160	121
U.S.A.	?	95	18
Norway	14	36	104

Source: *Quinn's Metal Handbook,* various editions.

In 1967 Brubaker singled out the United Kingdom as the outstanding example of lack of self-sufficiency in aluminium: 'The United Kingdom is anxious to get its metal at low cost and seemingly has no aspirations to produce more of its own.'[4] Almost immediately afterwards a volte-face in national policy set Britain off on a more than seven-fold increase in capacity at an overall cost of almost £200 million. The motivating forces were government decisions on imports and on energy policies. In the early sixties consumption of aluminium was steadily increasing so that by 1966–7 Britain was a £120 million a year finished aluminium market. Large import bills for the metal contributed to Britain's balance of payments difficulties. By the late sixties world surplus capacity in primary aluminium had disappeared, while consumption was growing rapidly. Looking ahead to 1975, it was estimated that British annual consumption might reach 600,000 tons with a consequent compounding of the import bill. Three other factors contributed to make it seem realistic to consider new British smelters. Prospects for new low power costs were of prime importance. Another factor was the desire to put the industrial house in order before joining the Common Market where France and Federal Germany produced 361,000 tons and 253,000 tons aluminium respectively in 1967. Finally, the semi-fabricators – dominated by four big groups, British Aluminium, Alcan, Impalco and James Booth – were suffering from over-capacity and there was a fear that if one of them built a home reduction plant it would be able to cut its costs and extend its fabricating at the expense of the others.

A rough-and-ready rule in the aluminium industry is that a power cost of less than 4.0 mills per kWh. is essential for economic reduction. In the mid 1960s Norwegian costs were well below this level. In terms of pence per electrical unit they were 0.3 to 0.4 pence at a time when the U.K. bulk tariff was 0.8 to 1.0 pence.

Table 32: Power cost to Norwegian reduction plants mid 1960s
(mills per kWh)

Own power 1.33	For newly built units 2.1–2.8
Purchased power 1.82	Fully amortized units 1.0

However, the Advanced Gas-Cooled nuclear reactor (A.G.R.) was expected to produce a new low in British electricity costs, 0.5 pence being quoted. This is still on the high side but low enough to make home reduction a marginally attractive proposition. Transport cost savings on ingots as compared with imports, and new and generous financial assistance to new industry in the Development Areas cancelled out much of the remaining cost disadvantage as compared with foreign supplies. It has been suggested that the 40 per cent plant and machinery and 25 per cent building grant which were available to firms in development areas were equivalent to a power cost reduction of 0.15 pence per unit, and that each 0.1 pence reduction cuts the ingot cost three per cent. Aluminium

199

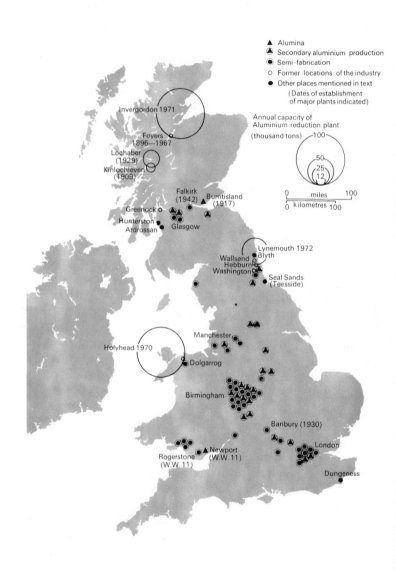

Figure 32   The British Aluminium Industry 1970–71

reduction in Britain thereby becomes economically attractive to the private concern, though not necessarily to the nation.

In 1955 associates of the group of mining and industrial companies now known as Rio Tinto Zinc discovered major bauxite deposits at Weipa, Australia. Apart from sales to independent aluminium producers R.T.Z. had only the relatively small Bell Bay, Tasmania, reduction plant. Its share of British semi-fabricating was slight, essentially the Widnes extrusion plant. In 1962 R.T.Z. suggested the possibility of a coal-based British smelter, but the N.C.B. declined the contract. By 1965 the Central Electricity Generating Board was beginning to talk of the prospect of power at 0.5 pence per unit from its Dungeness 'B' nuclear power station. R.T.Z., therefore, proposed that it should help finance a new nuclear station in return for cheap power for aluminium reduction. The Department of Economic Affairs refused the very large capital assistance needed for the power station.

Spurred on by R.T.Z.'s initiative, Alcan and British Aluminium made contact with the Highlands and Islands Development Board in 1966 and early 1967. By mid year all three companies had submitted detailed schemes. Pressure from the Scottish Office ensured that the government was willing to consider the possibility of more than one British plant. In the following month the issue became a matter of party politics when, as reported by the *Economist*, a '. . . large-scale aluminium smelting industry was conjured up by the Prime Minister before a dutifully spellbound party conference audience'.[5] By this time five or six companies were interested and the necessary formula for the provision of cheap power to them was taking shape. Although they benefited from bulk supply tariffs, other manufacturing industries paid a price which included some cover to the capital charges on the whole electricity generating and distribution system. Under the new plan the smelters would pay merely for the capital cost of the generating capacity which they would need, and a unit price reflecting their very high load factor.

Four smelters had been proposed by early 1968, though the government had committed itself for special help to only two. British Aluminium and Rio Tinto Zinc were regarded as the most likely successful candidates. Each proposed a 120,000-ton smelter with purchase of electricity along the lines outlined above. Alcan and Alusuisse had submitted plans for smelters of 60,000 tons rising to 120,000 by 1974, based either on nuclear power or alternatively on coal purchased for use in their own generating stations.

The Coal Mines Nationalization Act of 1946 had committed the N.C.B. to supply coal without 'undue preference'. In 1967 the Central Electricity Generating Board paid an average price of five pence a therm for its purchases of 67 million tons of coal, but now the N.C.B. offered the aluminium companies 3.25 pence a therm falling to 3.0 pence. However, a powerful aid to the Coal Board power case was provided by the White Paper on Fuel Policy in the autumn of 1967. This involved a switch of national policy to a freer competition in the energy market, with coal losing at least some of its privileged position. A sharp run down in mine employment into the mid 1970s was anticipated. As a result the prospect of new assured coal outlets was extremely attractive, not only to the N.C.B., but also to the government. Moreover, as another benefit, Alcan pointed out that a coal-based smelter would involve a 30 per cent saving in capital cost as compared with nuclear power. When it published its proposals in January 1968, Alcan had already drafted a twenty-five year contract for coal supplies. These factors caused scant attention to be given to the Industrial Reorganization Corporation's recommendation that the Alcan and Alusuisse proposals should be rejected – apparently on the grounds that they were too small.

Early in 1968 location became a prominent element in the controversy. R.T.Z. had already spoken of Anglesey as a likely choice, but had also considered Ulster among other areas, and in the late summer of 1967 examined four Scottish sites, one possibly Ardrossan on the Firth of Clyde, one in Fife, Invergordon, and probably another Ayrshire site. By early 1968, however, it was planning actively for Anglesey. British Aluminium had opted for Invergordon and Alusuisse had commented that almost any site close to deep water was suitable, but it was assumed that a development area would be chosen. It now took a logical step and proposed that its U.K. subsidiary, Star Aluminium, should build a smelter on Teesside, well located to obtain coal from Durham pits.

Alcan was left in an anomalous position. Like British Aluminium it had proposed to build at Invergordon, but to use coal from the north-east coast. Whereas the north of Scotland Hydro-electric Board would sell power to B.A.C.'s Invergordon smelter for 0.467 pence per kWh. and the C.E.G.B. supply to R.T.Z. Anglesey for 0.450 pence, a coal-based station at Invergordon was not expected to get below 0.55 pence. It seemed clear that Alcan's proposal was marginal and the government was known to favour the B.A.C. and R.T.Z. projects. The Alcan power station and smelter at Invergordon were estimated as likely to cost £49 million, to be supplied by 40,000 ton colliers unloading at a pier and coal conveying installation costing perhaps £3 to £4 million more. In March 1968, in the middle of a competition for the Invergordon development with B.A.C., and in the course of a Dingwall public inquiry, Alcan announced that it had been decided that it should build instead at Lynemouth near Blyth, Northumberland.

The Blyth area is the centre of a number of large and low cost pits, above all of the jointly operated Lynemouth and Ellington pits. In spite of this, in 1967 the C.E.G.B. had paid as much as 4.8 pence a therm for its purchases from Blyth area mines. C.E.G.B. complaints that Alcan was receiving an unrealistically low-priced fuel were countered by the N.C.B. revealing that coal had been offered at 3 pence a therm for a new major coal-based generating station rather than the nuclear one which the C.E.G.B. proposed to build at Seaton Carew, Hartlepool. Star and British Aluminium had already considered and rejected a Blyth area location.

Blyth was attractive to a government already being pushed beyond its intention to support only two smelters. Location there would bring capital saving, some new employment and, above all, a boost in construction work and in status to a district likely to be severely hit by run down of mine work forces, and where less than eighteen months before 1,000 jobs had been lost by a major shipyard closure. The 'Miners Group' of Labour M.P.s had come out strongly in favour of a coal-based project, and as Alusuisse was fading from the scene this implied support for Alcan. The firm

203

27. *Lynemouth aluminium smelter.*
The smelter, greatly delayed by labour troubles, came into operation in
1972. Its power comes from the Alcan power station seen near the shore, which, in turn,
obtains coal from the pit, top left.

rescued a project which seemed in danger of being rejected; Blyth received an important image boosting development.

Alcan's proposed 60,000 ton plant represented the technically optimum size of potline. B.A.C. and R.T.Z. had decided to aim for other scale economies by duplicating this module, that is, each planned plants of 120,000 tons. Now, partly to moderate the force of Norwegian objections that a subsidized aluminium industry contravened the rules of the E.F.T.A. convention, and partly to accommodate Alcan, the Board of Trade persuaded them to reduce their initial capacity from 120,000 tons to 100,000 tons, that is to sub-optimal potline size.

In the middle of 1968 the C.E.G.B. dealt the same two projects another blow by raising the charge for power from 0.5 to 0.6 pence a unit. Even so, this price is still based on their share of the capital cost plus the low running cost of the two power stations with the lowest costs in Britain, the new units of Dungeness 'B' and Hunterston 'B' which only nominally will supply them. A further advantage of Alcan's Blyth smelter stems from its location in a Special Development Area. The Special Development Areas were created to meet the problem of critical sections of the development areas where problems of rapid rundown in coalfield employment were especially severe. In these the building grant was 35 per cent as compared with the 25 per cent usual in a development area. Finally, construction of the Alcan plant was first off the mark, but unofficial strikes subsequently set completion back at least six months. In December 1970 the R.T.Z. smelter at Holyhead came into production. B.A.C.'s Invergordon units were at work by early summer 1971 but Lynemouth was delayed still longer. The three smelters involve capital investment of about £150 million. Highly favourable government loans of £62 million plus investment grants of about £60 million represent an extremely large public commitment. On the other hand, it was estimated when the projects were approved that substitution of home-produced for imported ingot would save £40 million of foreign exchange annually. By autumn 1970 the annual import bill saving was put at £55 million with perhaps as much as £10 million more as foreign exchange income from export of ingot, though this will disappear as home demand grows. Further expansion is envisaged, with Alcan, Lynemouth, at present the smallest project, doubling to 120,000 by late 1972.

Smelters, Semi-Fabrication and Regional Economic Growth

Hopes have been expressed that the new reduction plants would not only provide new employment directly, but would encourage the establishment of local semi-fabricating. Soon after the Lynemouth location had been chosen, the Managing Director of Alcan Aluminium (United Kingdom) went on record as saying that past experience of smelters 'established in the wilderness' showed that they attracted an equal number of jobs in ancillary industries.[6] Presumably he was overlooking the fact that his company's smelter was being built in a well populated, old industrial area, not in a pioneer zone. However, local authority leaders and M.P.s were also expressing the hope or the conviction that the whole development would snowball. In fact the pattern of British semi-fabricating is already well established, the industry has considerable problems of over-capacity and is being rationalized, so that further wholly new fabricating operations near to the new smelters are unlikely.

In the mid 1960s, before the smelter projects were mooted, British aluminium concerns were already reshaping their fabricating operations. Alcan modernized its big mills at Rogerstone and Banbury and made extensions to rod capacity in Birmingham. British Aluminium had four rolling mills in 1962, but closed the oldest, that at the over seventy-year-old Milton site, in that year, and five years later switched Warrington over from rolling to extrusion work. Volume mill operations were concentrated at Falkirk with jobbing operations at Resolven. The only other major firms were Impalco and James Booth, with smaller operations conducted by Birmetal, E. and E. Kaye and Indolex.

At the end of 1968 Alcan and Kaiser (which has a 40 per cent interest in the Anglesey smelter) each took half shares in James Booth. This firm operated a 35,000-ton mill capacity at Kitts Green, Birmingham and Skelmersdale, Lancashire. The new company was named Alcan Booth. Meanwhile, I.C.I. sold its interest in Impalco, which was renamed after its other previous part owner, Alcoa of Great Britain. In 1970 Alcoa announced a new rolling mill costing some £20 million for its existing works near Neath. Later in 1970 Alcan Booth began a further rationalization programme which will make Rogerstone the group's heavy tonnage mill centre with Banbury concentrating on extrusion, Kitts Green is to emphasize heavy plate production and specialist rolling operations.

These developments are indicative of two important conditions. Firstly, there is so much over-capacity and need for rationalization in existing works that little new semi-fabricating plant will be needed for many years. In aluminium rolling, particularly, substantial scale economies apply, but the average British mill is far too small. Péchiney, the chief European producer, is installing mills to roll 200,000 tons or more a year, but the total British output of rolled products in 1968 was only 180,000 tons and was produced by ten mills. Even though the turn out of specialist mills is necessarily small and so depresses this average, the situation is clearly very unsatisfactory. The second principle is a logical extension from the first; there is little scope for a break away from the existing pattern of market orientated semi-fabricating in Britain. Only R.T.Z., with a new smelter but controlled outlets which can take less than the smelter's capacity, may need new semi-fabricating plant. Consideration of the likely impact of the Alcan, Lynemouth smelter on the northern region seems to indicate that winning the smelter was in some senses at least a hollow victory.

Stage One of the Lynemouth smelter, 60,000 tons of capacity by summer 1972 has required substantial imports of material. It has been decided that these will come through the port of Blyth which, though much nearer the smelter, cannot handle vessels as big as the River Tyne. Annual traffic will involve 115,000 tons of alumina, 25,000 tons of petroleum coke, 10,000 tons of fuel oil, 7,000 tons of pitch and 2,000 tons of fluoride. As the plant doubles through to Stage Two in 1972, so will this traffic. Blyth harbour is handling a dwindling coal export so that the smelter's raw materials will merely maintain the existing work force.

The average labour force during construction of the smelter was 1,500. In Stage One operations 472 people will work in the smelter and 100 in the power station. In Stage Two the respective figures will be 660 and 150. Much was made of the fact that the Lynemouth contract would guarantee the employment of 1,000 miners, but this workforce is in pits which would have had a long life in any case. In 1970 the N.C.B. unsuccessfully applied for planning permission for a very big new open-cast coal site at Butterwell, just to the north-east of Morpeth. It was pointed out that this was necessary in order to meet long-term contracts which would otherwise be endangered by diversion of Ellington and Lynemouth coal to the smelter. Employment at Butterwell would be only 500. In short, the net additional N.C.B. labour force which the smelter is supporting

207

is really 500, not 1,000. Other calls on the region are for services, but only to a small extent for materials. One interesting exception is the case of fluorspar. When Alcan decided on Lynemouth, they proposed to import 2,000 tons and later 4,000 tons annually from Mexico, the world's largest producer. They were then informed that County Durham produced 28,000 tons of fluorspar a year and put in contact with local firms.

The northern region consumes under 1 per cent of the national total of semi-fabricated aluminium; over three quarters of consumption is in the Midlands and the South-east. Moreover, the average ton mile cost on rolled products is about twice that on primary aluminium ingots, and the price of ingots is a delivered price rather than a f.o.b. one, so that there is no agglomerative force pulling the semi-fabricator, even if very slightly, in the direction of the smelter. In short, Lynemouth, and the other smelters, too, provide locally important direct and multiplier new employment. They 'warm up' the economic atmosphere by giving a very desirable visual impression of dynamism. That apart, they cannot be expected to become major agents in regional transformation.

10. Mining and Economic Growth

Mineral working frequently involves the opening of inhospitable areas, peoples them and then, when mining declines, leaves them desolate. The ghost town descendant of a prosperous mining era is a common enough character of the drama of the old wild west, but one need not go so far in order to see numerous examples of the same phenomena. In other cases the elimination of mining from a local or regional economy does not involve the collapse of the structure, even if mineral exploitation was the spark which set growth in train. One needs to look first at the process of economic growth in a simplified manner. Later, looking in turn at a new copper project in South Africa, mining in the interior of Western Australia and the role of gold in the economy of the Rand and of the Republic of South Africa, the richness of the variations on the theme of economic growth associated with mineral working may be seen.

The holidaymaker from the crowded August beaches in Cornwall need only travel a mile or so inland to see the wastelands of a former major mineral economy built upon the exploitation of tin and copper ores. Until the middle of the nineteenth century workers were flooding into the mine villages of this area. Bigger centres such as Camborne and Redruth grew to serve the needs of mining companies and mine workers. Now, depleted of their old functions, their shopping centres bravely try to serve the smaller coastal resorts, while their suburbs straggle out along the main roads amidst the ruined engine houses, the old waste heaps and the abandoned mineral railways – the wilderness of their former glory. The population of the area is still too big for its present resources and, although there are other contributing factors, unemployment levels are high by national standards. Here in microcosm can be seen the local impact of growth and decline in a mineral economy – albeit one of early date and therefore of a widely scattered activity which contrasts sharply with the centralized operations common in the bigger mineral fields of the world today.

In other old British metalliferous mining districts and particularly in the lead dales of the northern Pennines or the former lead district at Shelve near the Stiperstones on the Welsh border the same story is repeated. On a very much bigger scale the manufacturing districts which sprang up to use local coal and iron ore are examples of the same thing, but here, in the nature of the materials involved, economic growth proceeded much further. Significantly these are the centres of our persistent development area problem. Here, too, though there are some great agglomerations of people, there are also, in the least accessible parts, literal ghost towns. Travelling through these areas of Britain, preferably with a topographical and geological map, the visitor will see much that is fascinating as well as a great deal which is distressing. He will begin to appreciate some of the ramifications of mineral working. Elsewhere in the world, where mineral surveys are now being carried out or where the first shafts are being sunk, we may be seeing the preliminary stages in the evolution of either a great industrial growth area of the year 2,000 or of a problem district of the mid twenty-first century.

In the past, when minerals had been proved, there typically occurred a great influx of hardy, hopeful individuals – a pattern of which the most famous example is the 'gold rush'. The individual filed the claim, provided the labour and the capital – which was frequently no more than his tools for the daily stint of panning. The miner took his own gold to the assay office. Yet even from an early stage the small mining town provided essential services for the miners – non-mining activities in an area whose sole reason for development consisted in its mineral wealth. One need only follow in the mind's eye the course of the typical street in countless frontier films to be aware of the array of services which was provided even at this primitive stage of development. For the chief town of a mining district growth even then could be much more impressive. Following the secularization of its Spanish mission,

San Francisco declined, and in 1847 had only 450 inhabitants. In September of that year placer gold was proved in the hinterland of Sacramento, and soon much of the population of San Francisco had hurried to the gold workings. By 1860 the city's population was over 55,000, and it was the major transport, commercial and manufacturing centre of the far west.

At the beginning of a mineral area's development it may be necessary to bring food in from outside. In some very inhospitable mining areas, as in northern Chile, it must always come from elsewhere, but in rather less extreme conditions the miners will grow some of their own supplies, or, more commonly, others will respond to the opportunities presented by the mining camps and begin to farm. In old British mining areas fields once cultivated by the workers may still be made out lying between the land still farmed and the unimproved moorland, although they are now invaded by bracken and surrounded by broken walls. In arid Nevada the population went up and down following a course which closely mirrored the oscillations of mining prosperity. The Californian gold rush was accompanied by the opening of large new crop and livestock areas in the great central valley, and in the nineties and first years of the twentieth century, at the time of the Yukon gold rushes, the cultivated area in that Canadian province reached levels far in excess of any approached since. In this case the regional economy declined as mining contracted or was centralized, but in other instances, once triggered off, farming gained a momentum of its own. After the impetus derived from the finds of 1849 and subsequent years, Californian agriculture became concerned with a very large export trade in wheat in the middle decades of the nineteenth century. The cattle economy of the northern Great Plains much further east began in the early 1860s with herds brought in from Texas for rearing before being slaughtered to supply the mining camps of Colorado, but in the seventies and eighties the stock-rearing glory of the Cattle Kingdom years had become an autonomous commercial activity. After the gold and silver rushes of the American West another aspect of mining was also illustrated there – the spread from one type of mineral production to another: Disillusioned miners discovered other ores and sometimes had the

211

acumen to realize that, although the glamour of lead or copper was less than that of gold, its profitability might be every bit as great.

By the time some local food production had developed and shops and services were established for miners, and also for the farmers and stock-rearers and the other people of the town, mine output might well be sufficiently large to support other industries. Preparation of the ore before dispatch for refining in older mining districts would be an early development. Stamping and washing plant, later facilities for chemical separation and perhaps also smelting developed near the mine. These, like the engine house, would need fuel, and would eventually help to justify a more substantial transport link than the ox train of pioneering days. A branch railway would now be called for, or, if no railhead was near enough to provide this, the whole local economy might fail to grow any further and eventually wither away as other favoured districts progressed. As the mine and its ancillary plants developed, so maintenance work or the need for new equipment would make calls on distant firms or on the ingenuity of a jack-of-all-trades miner, but eventually in an important mining area a foundry or mine engineering plant might be built. To this the railway would deliver coal, foundry iron or perhaps steel and, if the plant's products proved of high quality, a regional 'export' trade in machinery might join that in minerals.

In these and other ways the mine economy snowballs as economic activity widens from its initial very narrow base. If the mine closes after only a short period the whole community may collapse, but at a more advanced stage it may survive the rude shock and be able to make the necessary adjustments. In the twentieth century, with a maturing social conscience, government help is frequently provided to rehabilitate a declining economy or to bolster a threatened one, though the amount and form of help depends partly on the national ethos and partly on the size of the community involved in the crisis.

This snowball growth process is the result of a regional 'multiplier' effect, the needs or 'inputs' and the products or 'outputs' of one basic activity causing the growth of other and therefore derived functions. By the same process these in turn stimulate further growth. Although the process is cumulative, there are leaks in the money flows which connect the various parts of the developing regional economic structure. Even at the simplest, pioneering stage of a mine economy not all the money made in the diggings was spent in the local store or saloon, and later some of the equipment, or the explosives, or the fuel would be purchased from distant suppliers. The multiplier effect of a localized activity is not only regional, it not only builds up a local economy and society but has national or international implications too. As was remarked a few years ago, every time a mineral strike is made in the Canadian northlands the cash registers begin to ring in the big industrial cities of southern Ontario, and, one might have added, in the U.S.A. and Western Europe too. The multiplier has always had these widespread effects. No doubt the lead mined and smelted in the Mendips supported fabricators and bath builders in southern Gaul or Rome. In the eighteenth and nineteenth centuries the copper of Cornwall and later that of Latin America was smelted in the Swansea area, while the related brass trades were growing in the west Midlands. Along with the trails and the railways, the ports and the miserable mining camps of Peru or Chile, the copper trade built up the fortunes of such immigrant families to South Wales as the Vivians, and so contributed to further investment and economic growth in the coalfield economy, and, incidentally, to the desolation of the environment by smelter fumes in those eastern approaches to Swansea now being rehabilitated by the Lower Swansea Valley Project. Occasionally the multiplier took even more peculiar forms in spreading the impact of mining.

A number of the aspects of mining and economic growth discussed so far may perhaps most conveniently be summarized by a recent example – the development of the Phalaborwa mine area just north of the Olifants River and on the edge of Kruger National Park in north-eastern Transvaal. Until 1951 this was bare open veld, though there was evidence of ancient mining and smelting. A government-encouraged project for national phosphate production led at that time to the opening of apatite deposits, and by the mid 1960s there was a major open-pit operation, a town of 3,500 whites and a model Bantu township as well. Vermiculite production began later in the same area as a response to the rapid growth in demand for this insulating material. In the late 1950s further geological surveys in the same area proved very large deposits of low grade copper ores and magnetite. The Palabora Mining Company built a pilot plant to work the ore in 1961 and after that began development work on the open pit which will eventually become one of the world's biggest copper producers. Production began in 1966.

Development of the copper project was expected to cost R74 million or, at the pre-November 1967 rate of exchange, £37 million, all but about £8.5 million coming from overseas sources. A refinery has also been built. Yet the money was in the main spent in South Africa, whose firms provided most of the equipment and the 24,000 tons of cement and 20,000 tons of structural steel. Through these industries the multiplier effect of the development of the open pit and smelter went rippling outwards through the economy of the country. By the summer of 1963 a railway had been built at a cost of £2 million to link Phalaborwa to the main network, and the Electricity Supply Commission was building a £3.2 million transmission line from its new power station near Witbank, 140 miles to the south-west. In all these ancillaries of the project there are implied further rounds of demands – of steel, and cable and of coal for the construction work and operation of the power link for instance – and more wage payments which indirectly stem from the decision to open the copper ores underlying this lonely veld site.

In 1966, with the Palabora Mining Company's operations underway, South African copper production more than doubled and export earnings from copper went up from £20.5 million to £46 million. Not only does Palabora contribute foreign exchange income in this way but it will also reduce the Republic's imports of highly refined copper. By 1969 Palabora also produced 2.4 million tons of magnetite annually. A final aspect of the project concerns its role in the country's regional economic planning policies. Since 1960 the South African government has operated a Border Area Industrial Growth Point policy, designed to boost economic growth away from the traditional Rand growth centre, and so to lessen the impact of continuing Bantu inward movement to find employment in already overcrowded urban areas. Under this Border Area policy loans, investment allowances, tax concessions and subsidized services are available. The Palabora project is in such a growth area, benefits accordingly and contributes in turn to the social as well as economic well-being of the country.[1]

The development of iron ore deposits in Western Australia provides an even more spectacular example of mineral-based snowball growth – spectacular in terms of scale and rapidity of development and also in relation to the way in which old economic values have been changed both at home and at distances of many thousands of miles.[2]

In 1938, after a period as a relatively small-scale exporter, the Australian government put an embargo on further overseas sales from Iron Knob and other orefields in South Australia. It was feared that the depletion rate might put the country's own future steel-making prospects at hazard. In 1951 a new orefield was opened up in Yampi Sound in the far north of Western Australia, but as late as 1957 the Minister for National Development described the long-term outlook for iron ore as 'most alarming', and suggested the reserves were sufficient only for forty years. However at the end of 1960, to encourage more active exploration, the export embargo was lifted, with astonishing results. By late 1967 proved reserves of iron ore of over 50 per cent metal content were already 15,000 million tons or over 40 times the level of ten years earlier.

Capital investment in Australian iron ore has been on a very big scale and the funds have largely come from foreign sources. For example, Kaiser Steel Corporation of California has a 36 per cent share in Hamersley Iron Proprietary Ltd, which, over a period of twenty months, spent £100 million in opening the ore body at Mount Tom Price, equipping the mine, and building a 182-mile railway to the deep water ore docks at Dampier. An obvious effect is to contribute to the already fearsome competitiveness of Japanese steel plants which dominate the market. By 1965 freight charges on Australian ore to Japanese ports were already about $4 a ton less than those from Brazil, which in that year supplied 915,000 tons. Australian shipments to Japan in 1966 were almost ten times as great as those of 1965, and from 1966 to 1970 increased from 2.0 to 36.6 million tons.

Mineral Resources

The Western Australian government pressed the ore mining groups to consider processing some of the ore in the country. Hamersley committed itself to a three-stage development – ore exports, a large pellet making plant, and eventually, though perhaps not until the 1980s, an integrated iron and steel plant. Such a works it has been suggested will be mainly concerned with the export of steel, but the Broken Hill Proprietary Company which has been making steel for over fifty years from South Australian ores has realized it must protect its interests in the west. By 1970 it seemed possible that joint Australian–American interests might soon build a steel plant in Western Australia.

Iron ore shipments to Europe began in 1967 although hauls over this distance can only be economical if undertaken in very large carriers. Within a very few years the availability of Australian ore carried in 100,000-ton vessels will be a further contributor to the emphasis on deep-water steelmaking sites in Western Europe. At present most British works must obtain Australian ore by transshipment from one of the deeper terminals in Europe, notably Rotterdam. Looking ahead perhaps twenty years it is already suggested that it may be economical to bring iron or even semi-finished steel from works at such locations as Dampier to be finished in Britain, and in other European countries. There will clearly be serious arguments against dependence upon such a development, some of them with a political basis, but the possibility must certainly be envisaged. Things have already moved a long way since the coal and iron fields of Western Europe were reckoned to give it a natural pre-eminence in iron and steel manufacture.

Phalaborwa and the iron ranges of Western Australia are both examples of development from a negligible pre-existing economic base, but what of their long-term future? If one tries to look ahead beyond the quarter of a century in which the Palabora Mining Company is expected to operate its open pit, beyond the possibility of shaft driving to the highest grades of ore below the feasible limit of pit working, what will happen to the area when mining eventually stops? Will there by then be a Phalaborwa industrial complex able to grow further on its own as some of the literature about the project has suggested? The life of the Western Australia ore fields

216

will probably be much longer, but there too the problem of rising costs will eventually become serious. If diversification has proceeded well, a local or regional economy may none the less survive the decline of the mining activity which first set it off. There are no better examples of this than the Rand, where gold mining seems to be on the decline.

Until the last third of the nineteenth century South Africa had a predominantly pastoral economy. Incensed by the restrictions of British control after the Napoleonic Wars, the fiercely independent Boer farmers spread out from the fertile Cape Flats on to the dry Karoo and beyond across the great plateau. By about 1840 they had penetrated as far as the northern edge of the High Veld, the rather infertile divide between the basins of the rivers Vaal and Limpopo, known as the Witwatersrand. Away to the north-east, gold was discovered in the valley of the Crocodile River in 1853 and at Lydenburg in 1873, but neither of these finds was of much significance in the international tally of gold rushes. Discovery of diamonds far away to the south at Kimberley in 1866 proved to be of much greater importance. Foreigners flooded into the diamond fields which were opened in a host of small operations. However these proved unable to tackle the problems associated with the deeper shafts which were soon needed, and along with a threat to the price from over-production, this made some central, rationalizing force essential. The capital and entrepreneurship for this emerged in the form of the De Beers Mining Company. In 1873 there were no more than 300 miles of railway in South Africa, but to serve the diamond fields the Cape government built a line to Kimberley and this was opened in 1855. Mining skills, enterprise, financial resources and transport facilities already developed for two thirds of the way from Cape Town provided the ideal setting for the discovery of the gold resources of the Witwatersrand in March 1886. The gold bearing quartzite or 'banket' was discovered in the range of hills which is the surface expression of the seventy-mile east–west outcrop continuing beyond in a wide arc.

Outcrop working was employed at first but it was soon necessary to resort to mining. This, and the early need for crushing operations and the use of the relatively new cyanide process for extracting the gold, made the Rand gold industry capital-intensive. Transport developments went ahead with great rapidity. Within three years there was a railway line along the outcrop, and by 1895 Johannesburg was already linked to four ports, Cape Town, Port Elizabeth, Durban and the Portuguese settlement of Lourenço Marques. The farmer was provided with a big new market and also with the possibility of joint use of transport facilities initially built to serve the mines, although on the other hand it was said that the emphasis on tapping the Rand led to neglect of railway construction in agricultural districts nearer the coast. Within a year Johannesburg had a population of 5,000 to 6,000 inhabitants but it quickly became something more than a mining camp. By 1896 the population was 48,000 and eight years later was said to be 150,000, a figure so large that Lippincott's, the authoritative American gazetteer, accustomed to recording the mushroom growth of towns at home, felt compelled to query it. Native labour was widely recruited, so that as early as 1892 35,000 Africans were in the gold mines and by 1900 the number was 97,000. Here indeed the process of regional economic growth based on mining took on unusual proportions and the implications went beyond the mining area until the Rand played a key role in national growth as well. This was a response to two conditions, the size and nature of the mineral resource involved, and the general backwardness of the country in which it was developed. Because of them, in a few decades Johannesburg had become the biggest agglomeration of economic activity in the whole of Africa.

Mining and quarrying provided 28 per cent of the gross domestic product of the Union in 1911, more than agriculture, forestry and manufacturing together. Gold mining held the central place with just under 70 per cent of all mine employment. Growing material needs in the mines, population increase, a central location in the country's railway network and the fortunate discovery of coal to the north-east at Witbank and to the south around Vereeniging helped to make the towns clustered around Johannesburg the industrial as well as mining centres of South Africa. Demand for explosives, chemicals for the smelters and the whole range of other mine 'stores' has grown and diversified. In 1965 only 5.4 per cent

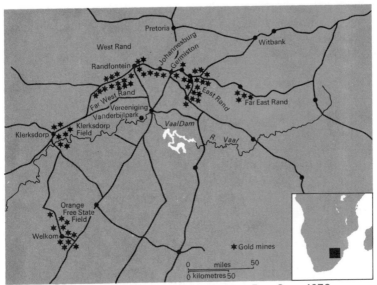

Figure 33   Gold Mines of the Rand and Orange Free State 1970

of stores for the gold mines were imported. Producers of other supplies for the mines and a host of consumer goods too congregated in the Rand area. Germiston, east of Johannesburg, became South Africa's chief railway equipment manufacturing and repair centre. At Vereeniging a small steel and rolling mill plant was established using local coal and Rand scrap, and in 1928, at the time of early national industrial promotion associated with the slogan 'South Africa First', an integrated steelworks was built at Pretoria by the South African Iron and Steel Industrial Corporation (I.S.C.O.R.). In steel development the central role of Rand gold is seen to perfection. The Pretoria steelworks are located between their coal supplies from Witbank and the iron ore derived from Thabazimbi in the north-west – or, more recently, also from Sishen far to the west in northern Cape Province. The nucleus of the railway network which links these together was a response to the needs of the Rand. From the Rand the works obtain their bought scrap, and as early as the end of its first decade of operations Pretoria annually sold steel valued at £6 million to the mines

219

alone. Further maturing of the Rand industrial structure has provided new outlets, and since the war another big integrated works has been built at Vereeniging, and in 1968 the country's third works came into production at Witbank. Each big new project adds to the region's labour force and purchasing power and therefore to future economic growth.

Gold has been the sheet anchor not only for the economy of the Rand but also that of the whole of South Africa. It has provided foreign exchange for the purchase of consumer goods and also to enable the country to bring in the capital equipment needed for industrial diversification. By 1966, although South Africa was already an important industrial country, gold exports were still almost 35 per cent of all exports. Immediately before the Second World War South Africa provided about 38 per cent of the non-communist world's gold, in 1958–60 it averaged 62 per cent and in 1969 77 per cent. It is acknowledged that the mineral discovery of March 1886 was the foundation stone for the regional and national growth of the last eighty years. Even so, it is arguable that now, with the decline in gold production expected, South Africa has an economy mature enough to stand the shock, and that the Rand's economic structure is sufficiently diverse to enable it to expand while its original basic activity withers.

In 1967 gold production fell away from the levels of the previous year for the first time since 1951. More significant than this, the slight increases that have been achieved in the last few years have come from beyond the traditional mining area, the stretch of ninety miles from Randfontein in the west to Nigel in the east. Since the war two new areas, known as the Far West Rand Gold-field and Far East Rand Goldfield, have been opened up, and together by 1963 they already had nine mines. Further south-west is the Klerksdorp field and in the Orange Free State west of Kroonstadt is the goldfield centred on the new city of Welkom, which began with the discovery of a reef in the autumn of 1946. There is no simple pattern of the concentration of high- or low-cost operations, as a consideration of the results of the widespread mines of the Anglo-American Corporation of South Africa will show, but broadly the outlying centres have lower costs and higher profits per ton of ore milled than those in the Randfontein–Nigel section. As early as 1963 the Orange Free State goldfield alone produced almost one third of South Africa's gold, and fields opened since the war together turned out 70 per cent. In the central Rand output has declined and some mines have closed.

The process of depletion and closure common to all extractive economic activity is complicated in the case of gold by the existence of a fixed price. After 1934 the world price of gold was stabilized at $35 per fine ounce, while both development and running costs increased. As a rough indication on the capital investment side, in 1933 it cost R5 million to float a new deep level mine but by 1966 for a comparable operation R39 million. Working costs in this time had increased from R2 per ton ore milled to R7.50. On the other hand, since 1952 revenue from sales of uranium by the gold mining companies has slowed the increase of costs borne by the gold tonnage. As a result the gold content per ton of ore necessary for a successful operation had in general risen from $5\frac{1}{2}$ penny weight (or dwt) to 10 pennyweight. With rising costs and a fixed price, closure of high cost mines became inevitable, low cost operations were opened to replace them, emphasis was placed on cost reduction and the government introduced a tax system which falls less heavily on the marginal mines. Even so, in 1967 the Gold Producers' Committee of the Chamber of Mines suggested that cost inflation in the mines since 1940 had caused the loss of production of 160 million ounces of gold which at 1967 exchange rates would have been worth £2,000 million. If this inflation went on at the current rate of 4 per cent a year against a fixed gold price well over one quarter of the gold workable at 1967 costs would not be mined. Already the rate of development has slackened so that, although £15 million has been spent in prospecting over the last ten years, the decision has been taken to develop only four more mines. In the mid sixties the Gold Producers' Committee suggested that production within twenty years might well be only one sixth the present level. This would mean a decline in foreign income of £300 million a year, or an amount equal to half the present level of all South Africa's other exports. In 1968 after a period of uncertainty the $35 per ounce price was reaffirmed for inter-governmental transactions but there is now no upper limit for non-monetary transactions. This raises returns on gold sales and has given a fillip to the mines, and production rose by over 3 per cent in 1970. In 1969 the free market price rose to almost $44. However gold's significance to the economy has already declined.

Gold mining contributed about 8 per cent to the Gross Domestic Product in 1965, twice as much as all other mining activities, but small in comparison with the 22.3 per cent from manufacturing. Between 1964 and 1967 the value of South African copper output went up 300 per cent, iron ore 70 per cent, manganese 68 per cent and chrome 40 per cent, while for gold the increase was 5 per cent. Employment in gold has fallen away from a 1961 average of 399,000 Africans and 49,000 Europeans to 368,000 and 40,000 respectively in 1968, and, as the mines have proportionally more Africans and fewer Europeans than other mining enterprises, their contribution to the purchasing power of the population is smaller than would be suggested by consideration of output values or the size of labour force. Moreover in 1969 over 65 per cent of African labour was recruited from outside the Republic so that its impact on South African purchasing power was still further reduced. For many years the relative contribution of gold mine salary and wage payments and consumption of stores in the mines has been decreasing and, although a sudden collapse would have disastrous implications, the country's economy will presumably accommodate the continuing gradual decline equally well. To the extent that it calls out the initiative necessary to open up other resources and to develop competitive manufacturing industry, the prospect of a slow decline in gold production may well prove a national blessing.

It has been suggested that subsidies might be paid to reduce the rate of decline so giving the rest of the economy more time to develop. Some other gold producers, notably Australia and Canada, subsidize their mines, but in these cases the output is very much smaller than that of South Africa and comes from widely scattered fields. Their subsidies were designed to maintain stable mining communities. South African conditions are very different, and the proposal for a subsidy has not been accepted. The increase in the price of gold on the free market permits the companies to lower the grade of ore worked and so lengthens the life of the mine. But, seeing the long term adverse trend, the great gold companies have diversified into other mining activities, to non-mining interests and other countries. For example, the giant Anglo-American Corporation of South Africa, while stressing its intention to retain its main interests in South Africa, in 1962 made an important investment in the Hudson Bay Mining and Smelting Company whose main concern is with copper and zinc operations in Manitoba and Saskatchewan but which is also involved in Baffin Island iron ore and in prospecting both in the U.S.A. and Canada. In the mid 1960s Anglo-American also formed Charter Consolidated Ltd as an investment company largely concerned with making available the Corporation's mining skills and expertise. In the spring of 1968

28. *Welkom gold district, Orange Free State, 1949.*
The first shops at Welkom, in the middle of the bare veldt, December 1949.

223

29. Welkom, Orange Free State today.
The suburb of Doorn.

this subsidiary became associated with I.C.I. in the opening of the potash deposits of the Staithes area of Yorkshire. Early in 1965 Anglo-American made the first deliveries of iron ore from its Swaziland iron ore fields to Japan, and at the same time decided to invest £35 million in the new Highveld steel and vanadium plant at Witbank. In the centre of Johannesburg it was already entering the field of real estate development on a big scale, joining with South African Breweries Ltd in a five-block development of shops, offices, apartments and an international class hotel.[3]

224

The ore reef on the farm of Welkom has reared a city of 100,000 in only twenty years, in eighty the discoveries of 1886 have created a city of over one million and a conurbation very much larger still. In 1969 South African gold production was worth almost £500 million but the mining companies are now facing the long-term prospect of decline and abandonment. However, it must be remembered that their wage payments have changed the economy and society of remote villages throughout southern Africa from which their labour has been drawn, their demand for stores has built up important factory trades, and their profits now go increasingly into new lines of economic activity and to some extent into other countries. In the light of this, to say that minerals are wasting assets, though true enough, is to miss their vital contribution to economic growth.

11. Mining and Planning

Like all economic activities mining is concerned with much more than merely its companies or its workers. As shown in the last chapter, it shapes patterns of regional and even of national economic growth. It also affects the physical landscape, through both the characteristics and what is now increasingly termed the quality of the environment. The scope for clash of interests is great and becomes greater.

Wealthy societies are beginning to turn from preoccupation with the growth of their G.N.P. to pay more attention to what is comprehensively called the quality of life. European Conservation Year 1970 brought this issue to a wider public although some of the events of autumn 1970 in Britain called in doubt the seriousness of our intentions. Yet on the other hand balance of payments difficulties encourage more home mineral working or processing, perhaps with highly undesirable landscape effects. Meanwhile the Third World countries, though not ignoring environmental planning, realize the desirability of developing and further processing more of their own natural resources in order to increase foreign exchange earnings or to lessen their import bills.

Ill-considered or badly constructed mining and mineral processing may litter the landscape with unsightly structures, solid waste heaps and transport facilities, pollute air, water and soil and kill off vegetation. As leisure increases in the affluent societies of the West, so the radius within which an increasing proportion of the urban population seeks it widens and regional ambitions for new employment may clash with the desires of visitors. At the same time scale of operation is increasing, there has been a strong swing to open-pit mining, and smelters grow bigger. At all levels there is scope for conflict over land uses and dispute about priorities. The question of planning for mining involves international, nationʼal and regional economic policies, private and social values, legal and statutory controls.

Legislation is important in the location and physical form of mining. It may lay down guide-lines, provide subsidies or build up a wholly new national mining enterprise. In the period 1938–45 the problems of ironstone working in the East Midlands of England were examined by a number of committees. Their report was followed in 1951 by the Mineral Workings Act which made restoration obligatory wherever possible. A Restoration Fund was set up with contributions from mining companies, landowners and the Exchequer. One effect was to make home iron ore mining marginally less attractive and to that extent to encourage overseas mining.

Regulation of atmospheric pollution affects mining and process-ing. The smoke control ordinances of American cities three quarters of a century ago greatly increased the consumption of anthracite and so boosted the growth of the distinctive Pennsyl-vanian anthracite economic region centred on Scranton, Wilkes-Barre and Hazleton. Current concern over emission of car exhaust gases may also have a long-term effect on mining. Anti-knock compounds are major pollutant makers and at the same time a major outlet for lead. The costs of altering engine structure are so great that in spite of forthright American action no revolutionary change may be expected.

Smelter operations have long been recognized as an environ-mental scourge in need of regulation. One hundred years ago there were fifteen copper smelters in the Swansea area with an annual output of £3.5 million. Their impact on the landscape contributed to the desolation, part of which the Lower Swansea Valley Project is now struggling to repair. In the U.S.A. copper smelting at Ducktown in the extreme south-east of Tennessee was notorious for its effect on the vegetation of the surrounding Appalachian slopes. Not until 1898, fifty years after operations began, was a sulphuric acid plant built there to make use of the wastes from the sulphide ores. The lead-zinc smelter at Trail in British Columbia is almost on the boundary with the U.S. state of Washington, so that its operations gained especial notoriety, pollution having inter-national implications.

Japan is currently experiencing acute difficulties with smelter effluents. These problems stem from the rapid growth of mineral consumption, the need to import a very large percentage of supplies – in spite of Japan's still considerable copper and zinc output – and the location of these smelters on the east coast where, because of Japan's topography, the great agglomerations of population are also found. By 1970 conditions were such that new smelter sites could only be found with difficulty and to control pollution some existing plants were having to reduce output, in one or two cases by as much as 30 to 40 per cent.

Most mineral extraction takes place well away from great centres of population though, as noted above, newly acquired mobility means that more of the population is affected. On the other hand widespread, low grade minerals, unable to stand long distance transport, may have to be worked near to big urban areas. This is especially so with sand and gravel, demand for which rises rapidly for house and factory building, urban renewal and so on. In Britain 1,400 or so pits produce about 100 million tons of sand and gravel annually: by 1980 a figure of 200 million tons is likely. The visual impact is distressing. Very large areas may be rendered desolate, and, although no very elaborate processing is involved, washing and separation is a noisy and sometimes a dust-creating operation.

In Britain the planning problems associated with sand and gravel working have been widely examined since before the end of the Second World War.[1] Some notorious cases came to prominence at that time. The most spectacular and complex involved aggregate working on the West Middlesex plain. Here land use conflicts were as sharp as can reasonably be conceived. The soils overlying the gravel supported intensive agriculture and market gardening. The spread of London caused a demand for building land and, at the same time, the decision was taken to build London airport at Heathrow in the same area. In short, in the quality of its surface and new pressures for development from above (aircraft), below (gravel) and laterally (the spread of London) the range of conflicts

Individual workings

Major built-up areas

0 miles 100

0 kilometres 100

Figure 34 Sand and Gravel Workings in Great Britain in the Mid 1950s

was complete, and at this critical time planning control proved too lax. Until 1947 no permission was needed before a quarry was opened and any sites opened before then remained outside planning control. Since 1952 it has been possible to require developers of new pits to restore the land afterwards. After they have been worked, old, unrestored 'dry' pits may become desolate tracts; 'wet' pits offer some valuable after-use possibilities, notably fishing and sailing.

West Middlesex is now an unhappy mixture of concurrent uses. Gravel workings cut up large sections of the plain. Arterial roads, housing and all the other material apparatus of an affluent society occupy much of the rest of the area. There are the remnants of high grade horticulture and farming, while overhead long distance jets drone into Heathrow at the rate of one a minute at peak times.

In few parts of Britain or the world is mineral working so obviously a contributor to a major land use crisis. However, as demand for aggregate grows rapidly pressures will mount in all densely populated areas. The Trent Valley, between the West Midlands conurbation and the incipient East Midlands conurbation is an example. The Solent growth area planned in the development strategy for the south-east is another, with the 1968 revival of an application to work sand and gravel under 600 acres of Hampshire pine forests a sign of the conflicts which are possible.

There are celebrated cases with other minerals too. The limestone quarries of the Pennines supply chemical and steel works, and cement works too. The biggest area of limestone working there is in the south, within or on the edge of the Peak District National Park. The outcry which greeted proposals to extend limestone working and cement making in the Hope Valley showed how deeply the intrusion was resented. In this case restoration is impossible, though eventually quarry faces weather to a rough conformity with the colour and landforms of nature. With other minerals restoration has long been attempted. Old coprolite operations in Cambridgeshire and East Anglia are largely restored, and almost ninety years ago restoration was actively underway in the small East Midlands ironstone workings. It was then usual to work the marlstone by driving a cut from one eighth to half a mile in length, after which

30. *Minerals and land use conflict. The West Middlesex Plain.*
The West Middlesex Plain, July 1959. Residential development, intensive
farming, Heathrow airport and numerous sand and gravel workings make up an
impressive conglomeration of land uses.

both or just one face was worked. The rubble was sifted by quarry
forks and the earth was spread over the worked ground. It was
claimed that the quality of the land was improved. 'We may
frequently see good crops of corn or grass growing on the made
ground to within a few feet of the present working face.'[2] Unfortunately the industry became mechanized and the result was the
creation of the notorious hill and dale topography, or the landscape
more colourfully described as the mountains of the moon. It was
this situation which made the Minerals Working Act of 1951 so
essential.

231

31. *Boulby potash mine, North Yorkshire.*
The mine, a joint venture of I.C.I. and Charter Consolidated, is located just within the North York Moors National Park. Great care is taken in landscaping.

Legislation and Mining in Britain

Institutional and legal conditions have been unfavourable to mining in twentieth-century Britain. Successive governments failed both to give the desirable low taxation rates for the initial period until 'risk' money could be recovered, and to take account of the long-term depreciation in mine values. There were high taxes on mineral royalties. Such harsh conditions have caused a number of concerns to draw back from mine development even when the results of prospecting and evaluation have otherwise been favourable. In marked contrast is the situation in Eire where spectacular developments have followed the liberalization of minerals legislation. Another difficulty stemmed from the fragmentation of British mining rights so that it was sometimes judged impossible to trace and assemble the titles. Nationalization of minerals or a central register of titles seems a way out of this difficulty. In 1969 the Labour government proposed new mining legislation to facilitate access to land and the acquisition of exploitation rights. Under this system a mineral owner would have no more than three months to object to an application for a mining licence and mineral rights could then be acquired compulsorily in the national interest. A clause was inserted into the 1970 Finance Bill to permit mineral owners to treat half their royalty receipts as capital instead of wholly as income. In the course of 1971 the Conservative government moved strongly to a position of support for more home mineral exploitation.

In addition to national legislation on rights, royalties, taxes and so on a mine developer has to overcome the hurdle of planning permission, the immediate decision resting with the local planning authority, and in mineral districts usually a County Council. Appeal against that decision can be made by either proposer or opponents to the Ministry of Housing and Local Government. How national and local influences and controls have worked and clashed and where the wider international issues come in must be traced.

In spite of a trend to easier government attitudes to mining in Britain, dispute about its desirability is acute. Mineral prices are volatile and most of them are rising. Overseas purchases put pressure on the balance of payments. In 1969 the Department of Economic Affairs showed that in the previous year Britain's mineral production was worth £943 million – mostly coal, iron ore, sand, gravel and limestone. Imports were more than £2,000 million, £300 million of which was for copper, zinc, lead and tin. With new mining legislation in the offing, the D.E.A. hoped for '. . . a substantial increase in mining activity and import saving . . .' The idea of the feasibility of renewed mining in British base-metal areas possesses a fascination owing something to romance as well as to economics. At first sight there would be other gains. A boost would be given to the local economy and still more to morale in some of the most marginal areas of our marginal regions. Unemployment rates are high in west Cornwall, west Durham and very high in Anglesey, yet in these same areas mining could benefit from development area legislation assistance which grew very rapidly after 1966.* To this case there are powerful opposition points.

It remains questionable whether a crowded, mobile society like that in Britain can afford the room and the affront to amenity which even the most cleverly disguised mining may constitute. It is of course easy to be obscurantist at this point, forgetting the great scars of past mining or of current coal and iron ore working, both of which are very much more extensive, and to remedy which relatively little has been done, and also minimizing the skill with which mining may now be landscaped. On the other hand landscaping increases costs and makes the economics of home mining more doubtful. Moreover, except for a few of the newer pits, coal mining is confined to areas which are already populous, but much of the proposed new non-ferrous mining would be remote from population and either in wholly new mineral districts or in areas which, though worked in the past, have sometimes had their old scars softened by time. The proposed £60 million R.T.Z. open cast copper workings in the Snowdonia National Park and the working of gold on the Mawddach estuary are cases which are sufficiently controversial and contemporary not to require further elaboration.

Moreover, if on environmental, conservationist grounds new mining is opposed in Britain, there are also strong economic arguments for ensuring that expansion should be located in the Third World. There mineral revenues may be the touchstone of development to countries which prefer trade to aid. It is true that in recent years the developing nations have been squeezing the more advanced countries, by moving towards more complete control over mining, bigger royalties and so on, but in large part this is a natural

---

* A recent example of the case for extension of mining in Britain is provided by two letters to *The Times*, 15 and 24 December 1970, from Sir John Lomax, formerly Ambassador to Bolivia. 'First things come first: being before wellbeing: preserve local resources for local labour before scenic values.'

Figure 35   Minerals and Environment in Great Britain and Eire 1968-71
Based on Mining Annual Review

reaction to the narrowness of the past views of responsibilities for economic advancement of raw material producers held by the industrial nations. It seems irrational to give aid and at the same time to reduce dependence on one of their most important sources of revenue. Revival of copper working on Parys mountain, a flicker of new life in the mines of the lead dales, may provide a few jobs but it will contribute, however marginally, to slow down the development of economies such as those of Zambia and Burma, whose respective per capita G.N.P.s in the mid 1960s were 11 per cent and 3.7 per cent that of the United Kingdom. The same applies to the preference for new British aluminium smelters rather than participation in the expansion of the Volta River project. A closer examination of one case may point up the issues more satisfactorily.

Long famous for copper and still more for tin, Cornwall was still among the world's leading producers a century ago. Fortunes were made from its mines. Thriving communities, roads, railways, ports, steamship routes and the innumerable other ancillaries and 'spin-off' of a major mining enterprise resulted. Over the town of Red-ruth rises the rough and impressive outcrop of Carn Brea. From the top a broad view of the former mining district, now desolate, can be seen. Three miles to the west was Dolcoath mine. Large-scale development there began in the early eighteenth century. Copper operations yielded about £5 million to shareholders, but after the 1830s tin became its chief product and by 1890 it had produced block tin worth £9 million. Along with the other mines Dolcoath transformed the area: one estimate suggests that by 1860 the population of the Camborne neighbourhood was already six times that of 1770. But by the 1890s Cornish mining was struggling with competitors and had too little capital to modernize. Dolcoath closed in 1920. The other mines failed one by one, until only two were left, South Crofty near Camborne and Geevor on the coast near St Just. Cornish miners migrated to other mining districts at home and overseas, carrying with them not only their skills but a host of distinctive cultural traits. The home district retained the Camborne School of Mines, which is still a major producer of mining engineers, and the important mining engineer-

32. Redruth tin and copper mining district 1904.
A view from the engine house of the Dolcoath mine September 1904. Carn Brea
is the monument-topped hill on the right. Note the intensity of the impact of a large
number of small operations on the landscape.

ing firm of Holmans with a world-wide reputation. The district unemployment rate is high, for no major new industry took the place of mining. For the rest, west Cornwall retains in dereliction the material traces of its former greatness.

There was periodic discussion about revival of Cornish mining, but each time interest flickered only to fade. Keen competition from overseas placer deposits was the chief problem. Even when it became clear that easily worked placers would be exhausted in a few decades, when tin prices rose and as modern pumping made possible working under conditions quite unacceptable at the time the mines closed, many problems remained. The absence of tax concessions on mining, and the lack of detailed overall survey in a mineral field which had been worked by a host of separate concerns, leaving inadequate records, were just two of these. As the Chairman of Geevor Tin put it in 1964, it was difficult to get mineral

237

rights from those who '. . . do not know the value of what lies under their ground'. A further problem was the need to overcome local opposition and to obtain planning permission. There was also doubt about the local availability of skilled labour, though the labour needs of modern mining are much smaller than when the mines were at their most prosperous.

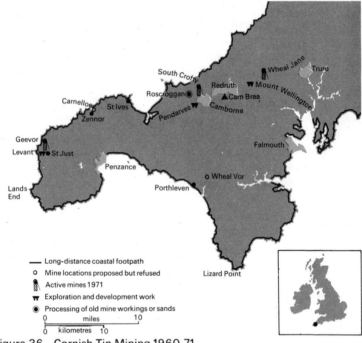

Figure 36   Cornish Tin Mining 1960-71

In 1961 application was made for permission to proceed to the reopening of the old Carnelloe mine at Zennor on the north coast of Penwith, a mine closed almost ninety years. Employment might be provided for 200. A public inquiry was held in Penzance. Zennor lies on a stretch of coast designated an area of outstanding natural beauty, and the proposal was opposed by the National Trust, the National Parks Commission, the Council for the Preservation of

238

Rural England, the Youth Hostels Association and other corporate and private interests. Application was refused, but in the mid 1960s there was further prospecting in the same area. Then in 1965 the Dartmoor National Parks Committee refused a request for permission to search for tin on Dartmoor from Consolidated Goldfields and emphasized that it was '. . . entirely opposed to any form of mining exploitation on the moor'. In the last year of the Conservative government the Minister of Housing and Local Government rejected a proposal to reopen Wheal Vor mine near Helston because it would deprive the area of a major water source, the mine supplying 300,000 gallons a day to the Helston and Porthleven Water Company. He stressed that he was not opposed to new mine development in principle.

After 1966 prospects improved. Financial aid to development areas mounted rapidly and tin prices were rising too, while some of the chief producers, notably Indonesia, Bolivia and Nigeria, were afflicted with a variety of disturbances. Geevor mine is expanding and South Crofty plans to double its output between 1969 and 1971. By 1969 at least seven companies were busy prospecting for tin in Cornwall, and in at least two other cases metal was being extracted from old mine tailings.

The first big new Cornish mine in fifty years was announced in 1969. Part of this development involves the reopening of the old Wheal Jane mine west of Truro. The capital cost is about £6 million, £2.4 million of which was to come from government sources as an investment grant. Wheal Jane will provide work for over 300 in an area where the winter peak unemployment rate is about three times the national average, and will produce some 150,000 tons ore yielding 1,300 tons of tin a year. This is equal to about 7 per cent of Britain's annual tin consumption and will save tin imports worth £2 million. By the late seventies a possible six operations could be producing almost 11,000 tons of Cornish tin – a level reached a century ago when 340 mines were at work.[3] This undoubted gain to the Cornish economy and the national balance of payments is not secured without other less desirable effects.

Wheal Jane's annual output will equal about 4.3 per cent of Bolivia's 1969 tin output. There, it would provide both work for

239

about 1,000 poor miners and vitally important export income to a country with a standard of living almost unbelievably low by our standards – 1965 income per head was £486 in the U.K. and £43 in Bolivia. Cornish tin development is indeed another case where politically induced economic developments in Britain, palliatives but not solutions to our balance of payments problems, go right against our once dearly held principles of the international division of labour, flout the 'trade-not-aid' pleas of the Third World and negate our expressions of solicitude for the advancement of poorer nations. Although the amenity issue can easily be overdone, it is ironic that one of the prices we shall pay for our petty, autarchic policy is some reduction in the attractiveness of one of the most popular holiday counties in Britain.

12. Mineral Resources for the Future

The growth of industrial economies led to a tremendous upward spiral of mineral consumption. This was accompanied by a shift of emphasis from precious or semi-precious metals to base metals and non-metallic minerals for building and for chemical purposes. After the great age of geographical discovery, the European nations pillaged the less developed world for their easily transportable, high value minerals. In the twentieth century, as the gulf in standards of living between classes within the advanced nations has narrowed, increasing flows of low grade minerals have contributed to a widening of the gap between the living standards not of classes but of groups of nations, between those of the advanced and of the Third World.

In the past it has been common for estimates of growth of mineral demand to be too conservative, and for American projections in particular to underestimate the speed with which accelerating consumption would spread to other industrial nations. This may be seen by comparing the 1975 projections of the Paley Report of 1952 with actual output levels by 1969. In the late sixties world annual consumption of aluminium and copper was 8 million and 5 million tons respectively. A. G. Charles of the British Metal Corporation has forecast that by the end of the century annual demand will be 70 million tons of aluminium and 20 million tons of copper.[1] In addition to growth of demand for existing major metals it is worth remembering that technical change may quite suddenly cause demand for new or minor metals. A century ago aluminium was a metallurgical curiosity used for jewellery or as an imperial substitute for gold and silver plate. Demand for uranium mushroomed in the forties and fifties for, as H. H. Read neatly put it, 'Uranium-bearing minerals once had an academic significance since they

Mineral Resources

Table 33: U.S. and non-Communist World Mineral Production and Consumption, 1950, 1969 and 1975 (million metric tons)

	1950	1975 (1952 estimate)	1969 (actual)
Copper			
U.S. production	0.8	0.7	1.4
consumption	1.6	2.3	?
Rest of non-Communist world			
production	1.4	2.7	3.4
consumption	1.2	1.8	?
Lead			
U.S. production	0.4	0.3	0.5
consumption	1.1	1.7	1.2
Rest of non-Communist world			
production	1.1	2.1	2.0
consumption	0.8	1.3	2.0
Zinc			
U.S. production	0.6	0.6	0.5
consumption	1.0	1.4	1.3
Rest of non-Communist world			
production	1.2	2.2	3.7
consumption	1.0	1.5	2.8
Tin			
U.S. production	–	–	–
consumption	0.08	0.11	0.06
Rest of non-Communist world			
production	0.15	0.17	0.18
consumption	0.07	0.10	0.12
Steel			
U.S. production	92	136	127
Rest of non-Communist world			
production	63	127	289

Sources: *Paley Report,* 1952 and *Mining Annual Review,* 1970.

could be used to date the earth; they have now become of a different significance and may be used to put an end to all our dates.'[2]

That supply to meet these demands is or is not adequate cannot be stated with any precision, for supply is a matter of resources, knowledge of and means to exploit them, and the word resource is itself only comprehensible in relation to technology. As demand increases and techniques of mineral processing become more sophisticated so it becomes economic to mine lower grade deposits, to work richer ores under more difficult conditions of environment, mining or access and to process poorer or more intractable ones. Moreover our knowledge of the mineral resources of the world as a whole is still quite inadequate. As a recent United Nations report put it:

> Twenty years ago it was commonly believed that there was no oil in the Sahara. There was also wide concern that there was going to be a shortage of iron ore. But vast discoveries of hitherto unsuspected mineral deposits have been made quite recently. The truth is that the surface of the earth has hardly been scratched. Satisfactory stratigraphic correlations between the different regions have not even been made.[3]

The evidence of recent years proves beyond doubt that major new ore bodies may still be found. Even so, by definition, minerals are wasting-assets, non-renewable earth resources. The implications are important for individual mineral districts, national economies and the world's mineral based economy.

In a world in which the nations are embarked on the Second Development Decade to raise the standard of living of underdeveloped areas, and when world population is still increasing by about 60 million a year, exploration, development and production may not be able to keep pace with demand. As long ago as 1958 an international group of mineral economists met in Paris under the chairmanship of André Siegfried to discuss the long term mineral supply situation. They concluded that rising population and per capita consumption would need $6,000 million a year of investment in addition to the outlay to maintain present production.

The case of copper is instructive. By the late 1960s India consumed approximately 80,000 metric tons a year, the U.S.A. in 1968 1,385,000. The late 1969 population of the United States was 201 million and of India 524 million. Even if Indian per capita consumption rose to no more than one quarter the American level, her needs would increase by about 820,000 metric tons or equal to almost 14 per cent of world output in 1969. If the needs of Africa, Latin America and China are computed on a similar basis the supply prospect looks extremely daunting. Consumption would be increased 3 million tons or more than 50 per cent above present levels. Looking ahead to a time when demand is so great that all mineral concentrations that could reasonably be considered ore deposits have been used up, Harrison Brown of the California Institute of Technology suggested that by then the mineral industry will have to be reshaped. Instead of using strongly localized resources it will be necessary to break up rocks and process them. Even then the operations would be located where the concentrations of desired elements are greatest so that the operation would represent an extreme deviation in degree rather than in kind from present day mining and processing, but, as McDivitt suggested, an important line of division between industries would then have disappeared – 'Mining will no longer exist as an industry, having been absorbed into the chemical industry'.[4] Before that stage is reached there are possibilities of tapping two other sources of minerals. One may be minerals from another planet or perhaps from the moon. Human progress has been so spectacular that it would be foolhardy to rule this out as a mere Wellsian fancy, but we are clearly many decades from the stage when it will be economically more desirable than the extreme case of processing earth rock that Brown postulated. Much more within our grasp is the opening of the mineral resources of the oceans. There, perhaps, lies the mining frontier of the future.

Notes

Preface

1. Feiss, J. W., *Minerals: Technology and Economic Development*, Harmondsworth, 1965, p. 108.
2. *Mining Engineering*, February, 1969; *Mining Annual Review*, 1969, p. 5.

Chapter 1. *Introduction*

1. Prescott, W. H., *History of the Conquest of Mexico and History of the Conquest of Peru*, New York, 1843, 1847 (p. 829 of Modern Library edition).
2. Rickard, T. A., *Man and Metal. A History of Mining in Relation to the Development of Civilisation*, 2 vols., New York and London, 1932.
3. Read, H. H., 'The Geologist as Historian', *Scientific Objectives*, London, 1952.
4. *The Times*, 2 January 1971.
5. Zimmerman, E. W., *World Resources and Industries*, New York, 1951, p. 423 has wise words on this point.
6. U.N. Science and Technology for Development, Report on the U.N. Conference on the Application of Science and Technology for the Benefit of the Less Developed Areas, vol. 2 *Natural Resources*, 1963.

Chapter 2. *Geological Considerations and Mineral Survey*

1. *Economist*, 7 March 1953, p. 685.
2. Riley, C. M., *Our Mineral Resources*, New York, 1959, p. 85.
3. *Mining Annual Review*, 1970, p. 141. See also 'World Mining', *The Times Supplement*, 16 October 1970.
4. James, C. H. and Khan, M. A., 'Mineral Exploration', *Mining Annual Review*, 1970, p. 135.
5. *Remarks on the Second Five Year Plan, Birmingham Memoranda on Russian Economic Conditions*. November 1934, p. 11. *U.N. Non Ferrous Metals in Underdeveloped Countries*, 1956, p. 57 f.n. Read, H. H. 'The Geologist as Historian', *Scientific Objectives*, 1952.

6. *Mining Annual Review,* 1971, p. 301.
7. *The Times,* 2 May 1969.
8. *Mining Annual Review,* 1970, p. 239.
9. Resources for the Future, *Resources,* May 1966.
10. 'Sharing the Bed', *Economist,* 8 August 1970, p. 69.

Chapter 3. *Mining: Some Aspects of its Economics*

1. An excellent overall view may be obtained from the composite map 'Minerals and Industries' in *Oxford Economic Atlas of the World, 1965.*
2. *The Times,* 2 May 1969.
3. *Canada Yearbook,* 1955, p. 24.
4. Whitmore, R. L., 'Riding the Minerals Boom', Inaugural lecture, 1968, University of Queensland, Brisbane.
5. Pryor, E. J., *Economics for the Mineral Engineer,* Oxford, 1958, p. 13.
6. *Mining Annual Review,* 1969, p. 163.
7. ibid, 1970, p. 155.
8. ibid, 1969, p. 169.
9. ibid, 1971, p. 11.
10. For details see *World Mining,* October 1969.
11. See U.N. *Non Ferrous Metals in Underdeveloped Countries,* 1956.
12. *Bank of New South Wales Review,* September 1967.
13. *Progress* 2, 1968, p. 166.
14. Prain, Sir R., 'The Responsibility of a Mining Industry to the Community'. Lecture to the Royal School of Mines, Rhodesian Selection Trust, Salisbury, 1957.
15. *Peruvian Times,* 17 April 1970, pp. 8–10.

Chapter 4. *Mineral Transport*

1. *Observer,* 16 June 1963.
2. 'The Tanzam Railway', *Standard Bank Review,* December 1970.

Chapter 5. *Mineral Processing – Aluminium*

1. Mangin, A., *The Earth and its Treasures,* London, 1875, p. 260.
2. 'Bauxite, Alumina and Aluminium', *Overseas Geological Survey,* 1962, p. 2.
3. ibid, 1962, p. 64.
4. 'To process or not to process', *Metal Bulletin,* 18 September 1970, pp. 17, 24.

5. 'Volta River Aluminium Scheme', H.M.S.O. Cd. 8702, 1952.
6. See especially Jackson, Sir R., 'The Volta River Project', *Progress* 4, 1964, pp. 146–61.
7. U.N. Studies in Economics of Industry 2. *Pre-Investment Data for the Aluminium Industry*, 1966.
8. *Bolsa Review* May 1969, August 1969, May 1970, July 1970 and *Peruvian Times,* 3 April 1970.

Chapter 6. *Copper*

1. Prain, Sir R., *The Development of the Copper Industry*, 1957, p. 31.
2. Levy, Y., 'Copper. Red Metal in Flux', *Federal Reserve Bank of San Francisco Report*, Monthly Review Supplement, 1968, p. 46.
3. *Economic Bulletin for Latin America*, XIV, 2, 1969.
4. Powelson, J. P., *Latin America. Today's Economic and Social Revolution*, New York, 1964, p. 128.
5. *Peruvian Times,* 7 April 1970, pp. 8–10.
6. Chairman's remarks, A.G.M. Union Minière, 26 June 1951.
7. *World Business*, 9 November 1967.
8. 'The Copper Bottom?' *Economist*, 7 November 1970, pp. 64–5.
9. *Metal Bulletin*, 25 September 1970, p. 16.

Chapter 7. *International Trade and Price Problems in Minerals*

1. For a review of the problem see Meade, J. E., 'International Commodity Agreements', *Lloyds Bank Review*, 1964.
2. *The Economist*, 27 August 1960, p. 825; *New York Times*, 1 April 1960.
3. Smyth, W., *Five Years in Siam*, 1898, quoted in Little, A., *The Far East*, 1905, p. 276.
4. Gunther, J., *Inside Latin America*, London, 1942, p. 180.
5. Fox, D. J., 'The Bolivian Tin Mining Industry: Some Geographical and Economic Problems', International Tin Council, March 1967, p. 4.
6. Quoted in ibid, p. 11.
7. *Economist*, 14 August 1965, p. 601.
8. U.S. Bureau of Mines *Minerals Yearbook*, 1967, Vol. 4, p. 133 'Area Reports'.

Chapter 8. *The Political Factor in Mineral Exploitation, I*

1. Shimkin, D. B., *Minerals: A Key to Soviet Power*, Harvard, 1952, p. 303.
2. Lyaschenko, P., *A History of the National Economy of Russia to the 1917 Revolution*, New York, 1949, pp. 526, 538.
3. Saushkin, *Lectures on Soviet Geography*, Oslo, 1956.
4. *The Russian Yearbook*, 1913, p. 184.
5. ibid, p. 702.
6. Savitsky, P. in Malevsky-Malevitch, P., Russia: U.S.S.R. *A Complete Handbook*, New York, 1933.
7. Mikhaylov, N., *Soviet Geography. The New Industrial and Economic Distributions of the U.S.S.R.*, London, 1935, pp. 18, 26.
8. Shimkin, D. B., op. cit., p. 98.
9. ibid, p. 17.
10. Connolly, V., *Beyond the Urals*, Oxford, 1967, p. 314.
11. Shabad, T., *Basic Industrial Resources of the U.S.S.R.*, New York and London, 1970, pp. 56, 273.
12. Loginov, V., 'Regional Problems of Technical Progress', *Soviet Geography: Review and Translation*, January 1970, pp. 47–56.
13. ibid, pp. 55–6.
14. Yakovets, Y. V., 'On the neutralisation of the effect of physical differences in mining through the use of a system of prices', *Soviet Geography: Review and Translation*, March 1968, pp. 203–11.
15. *Mining Annual Review*, June 1970, p. 416.
16. *World Mining*, October 1969.

Chapter 9. *The Political Factor, II*

1. Elliot, W. Y., et al., *International Control in the Non-Ferrous Metals*, New York, 1937, p. 224.
2. The background of B.A.C. developments is considered in Chilton, L. V., 'The Aluminium Industry in Scotland', *Scottish Geographical Magazine*, Vol. 66, 1950.
3. 'The Volta River Aluminium Scheme', H.M.S.O. Cd. 8702, London, 1952.
4. Brubaker, S., *Trends in the World Aluminium Industry*, Baltimore, 1967, pp. 135, 240.
5. *Economist*, 7 July 1967, p. 68.
6. *Guardian*, 12 June 1968.

Chapter 10. *Mining and Economic Growth*

1. On the Palabora Mining Company developments at Phalaborwa I am mainly indebted to the *Standard Bank Review*.
2. See also pp. 72–5 above.
3. On South African mining see the very full annual reports of the main mining houses, and the monthly and annual review of the Standard Bank Ltd. See also C. Strauss, 'The Impact of the Mining Industry on the Business Outlook of South Africa', *Standard Bank Review*, October 1967, and *Metal Bulletin*, special issue on South Africa, summer 1968.

Chapter 11. *Mining and Planning*

1. See, for example, *Geographical Journal*, Vol. CIV, nos. 5, 6, 1944.
2. Wilson, E., 'The Lias Limestone of Leicestershire as a source of iron', *Midland Naturalist*, 8, 1885, p. 125.
3. *Mining Annual Review*, 1971, p. 453. See also *Mining Journal*, 8 January 1971, pp. 18–19.

Chapter 12. *Mineral Resources for the Future*

1. *Economist*, 15 November 1969, pp. 58–9.
2. Read, H. H., 'The Geologist as Historian', *Scientific Objectives*, London, 1952.
3. U.N. *Natural Resources of Developing Countries. Investigation, Development and Rational Utilisation*, 1970, p. 12.
4. McDivitt, J. F., 'Prospecting for Mineral Ores in Europe', Organization for Economic Cooperation and Development, 1962, p. 9.

Further Reading

1. GENERAL

Chapter 1. *Introduction and General*

Riley, C. M., *Our Mineral Resources*, New York, 1959.
Flawn, P. T., *Mineral Resources*, Chicago, 1966.
Checkland, S. G., *The Mines of Tharsis*, London, 1967.
Rickard, T. A., *Men and Metals*. 2 vols., New York, 1932.
Journal and Statistical Sources:
  Mining Journal, London.
  World Mining, San Francisco.
  Metal Bulletin, London.
  Engineering and Mining Journal.
  Mining Engineering, New York.
  Mining Annual Review (published by *Mining Journal*, London).
  U.S. Bureau of Mines, *Minerals Yearbook*.
  U.S. Bureau of Mines, *Information Circular*.
  British Metals Corporation, *Annual Statistics*.

Chapter 2. *Geological Considerations and Mineral Survey*

U.N. *Proceedings of the U.N. Scientific Conference on the Conservation and Utilisation of Resources. 2. Mineral Resources*, New York, 1951.
U.N. *Science and Technology for Development, 2. Natural Resources*, 1963.
National Academy of Sciences and National Research Council, *Resources and Man. A Study and Recommendations*. San Francisco, 1969. Chapters 6 and 7.
Jones, W. R. *Minerals in Industry*, Harmondsworth, 1943, and later editions.
Alexander, W. and Street, A., *Metals in the Service of Man*, Harmondsworth, 1944 and later editions.

Chapter 3. *Mining: Some Aspects of its Economics*

Prain, Sir R. *The Economics of Modern Mining*, and *The Responsibility of*

250

a Mining Industry to the Community. Special University Lectures, Royal School of Mines, London, 1957.

Pryor, E. J., *Economics for the Mineral Engineer*, Oxford, 1958.

U.N. *Non-Ferrous Metals in Underdeveloped Countries*, 1956.

Bernstein, M. D., *The Mexican Mining Industry, 1890–1950. A Survey of Politics, Economics and Technology*. New York, 1964.

Chapter 4. *Mineral Transport*

Manners, G., *The Changing World Market for Iron Ore 1950–1980*, Baltimore, 1971.

Chapter 5. *Mineral Processing – Aluminium*

Brubaker, S., *Trends in the World Aluminium Industry*, Baltimore, 1967.

Huggins, H. D., *Aluminium in Changing Communities*, London, 1965.

'Aluminium, a World Survey,' *Metal Bulletin*, December 1963.

Chapter 6. *Copper*

Prain, Sir R., *The Development of the Copper Industry*, Special University Lecture. Royal School of Mines, London, 1957.

Smith, B. W., *The World's Great Copper Mines*, London, 1967.

Levy, Y., *Copper: Red Metal in Flux*. Federal Reserve Bank of San Francisco, 1968.

Pederson, L. R., *The Mining Industry of the Norte Chico, Chile*, Evanston, 1966.

Chapter 7. *International Trade and Price Problems in Minerals*

Hedges, E. S., *Tin in Social and Economic History*, London, 1964.

Robertson, W., *Report on the World Tin Position*, International Tin Council, London, 1965.

International Tin Council, *Statistical Yearbook*.

Fox, D. J., *The Bolivian Tin Mining Industry: Some Geographical and Economic Problems*, London, 1967.

Yip Yat Hoong, *The Development of the Tin Mining Industry of Malaya*, Kuala Lumpur, 1969.

Mineral Resources

Chapter 8. *The Political Factor in Mineral Exploitation*

Shimkin, D. B., *Minerals. A Key to Soviet Power,* Cambridge, 1953.
Shabad, T., *Basic Industrial Resources of the U.S.S.R.,* New York, 1969.
Soviet Geography Review and Translation, American Geographical Society Monthly, New York.
Weekly Digest of the Soviet Press,
Connolly, V., *Beyond the Urals*, Oxford, 1967.

Chapter 10. *Mining and Economic Growth*

Pelletier, R. A., *Mineral Resources of South-Central Africa*, Cape Town, 1964.
Houghton, D. H., *The South African Economy*, Cape Town, 1967.

2. PERIODICALS

Aschmann, H., 'The Natural History of a Mine', *Economic Geography*, 46. 1970, pp. 172–89.
White, K. D., 'The birth of mining', *Optima**, December 1955, pp. 115–20.
Kennedy, W. Q., 'The scientific approach to mineral prospecting', *Optima*, December 1962, pp. 213–16.
Brock, B. B., 'A philosophy of mineral exploration', *Optima*, September 1960, pp. 143–58.
Brock, B. B., 'Modern mining practices', *Optima,* June 1961, pp. 100–112, September 1961, pp. 158–66.
Heath, K. C. G., 'New patterns of world mining', *Optima,* March 1969, pp. 15–31.

Baines, G., 'Potash harvest beneath the Canadian prairie', *Geographical Magazine*, XLI, 1969, pp. 814–18.
Thomas, T. M., 'Potash Mining in Saskatchewan', *Geography*, L, 1965, pp. 295–8.

Harrison Church, R., 'Port Etienne: a Mauritanian pioneer town', *Geographical Journal*, CXXVIII, December 1962, pp. 498–504.
Baker, G., 'Sleeping River Congo awakes', *Geographical Magazine,* XLII, 1971, pp. 778–85.

* *Optima* is the quarterly review of the Anglo-American, De Beers and Charter Consolidated companies.

Wilson, H. H., 'Australia's Modern Iron Age', *Geographical Magazine,* XLI, 1969, pp. 294–8.

Swindell, K., 'Iron Ore Mining in West Africa: Some Recent Developments in Guinea, Sierra Leone and Liberia', *Economic Geography,* 43, 1967, pp. 333–46.

Marshall, A., 'Iron Age in the Pilbara', *The Australian Geographer,* X, 1966–8, pp. 415–20.

Hilling, D., 'The Changing Economy of Gabon', *Geography,* XLVIII, 1963, pp. 155–65.

Swindell, K., 'Iron ore mining in Liberia', *Geography,* L, 1965, pp. 75–8.

Sneesby, G. W., 'Economic development in Swaziland', *Geography,* LIII, 1968, pp. 186–9.

Swindell, K., 'Industrialization in Guinea', *Geography,* LIV, 1969, pp. 456–8.

Heath, K. C. G., 'Making a desert give up its treasures' (Mauritania), *Optima,* June 1967, pp. 75–81.

Mawby, Sir Maurice, 'The way ahead for Australian mining', *Optima,* September 1971, pp. 125–34.

Mawby, Sir Maurice, 'Western Australia', *New Commonwealth,* July 1971, pp. 25–9.

Mawby, Sir Maurice, 'Rich iron ore deposits give Swaziland its long awaited railway', *Optima,* June 1964, pp. 84–7.

Mawby, Sir Maurice, 'Industrial awakening of Swaziland', *Optima,* December 1961, p. 221.

Schnellmann, G. A., 'Iron ore resources and the world's changing needs', *Optima,* June 1966, pp. 73–87.

Manners, G., 'Latter-day leviathans for ocean bulk transport', *Optima,* December 1968, pp. 166–73.

Craig, D., 'The Aluminium Industry in Australia', *Geographical Review,* L.1, 1961, pp. 21–46.

Driscoll, E. M., 'Weipa – a new Bauxite mining area in North Queensland', *Geography,* XLVII, 1962, pp. 309, 310.

Hilton, T. G., 'Akosombo dam and the Volta River project', *Geography,* L.1, 1966, pp. 251–4.

Griffiths, I. L., 'Zambian Coal: an example of strategic resource development', *Geographical Review,* LVIII, 1968, pp. 538–51.

Steel, R. W., 'Action for Tropical Africa', *Geographical Magazine,* XLII, 1970, pp. 7–15.

Siviour, G. R., 'Kilembe copper mines – Uganda's most important mineral deposit', *Geography,* LIV, 1969, pp. 88–92.

Mineral Resources

Etheredge, D. A., 'Zambianization on the Copper Belt', *Optima*, December 1969, pp. 182–7.

Rhokana Corporation, *Optima*, December 1951, p. 22.

Nchanga Consolidated, *Optima*, March 1956, pp. 32–4.

Nchanga Development, *Optima*, June 1963, pp. 84–5.

Nichols, C. P., 'The Bancroft mine – a tale of tribulation and triumph', *Optima*, September 1956, pp. 65–70.

Nichols, C. P., 'Bancroft mine in production', *Optima*, June 1957, p. 93.

Nkana-Kitwe: twin towns of copper, *Optima*, June 1961, pp. 68–9.

Broken Hill, Zambia, *Optima*, December 1962, pp. 192–3.

Focus on Zambia, *New Commonwealth*, September/October 1971, pp. 11–19.

Bennett, O. B., 'Large-scale mining methods on the Copper Belt', *Optima*, June 1953, pp. 19–24.

McTaggart, W. D., 'Developments in the New Caledonian nickel industry, *Geography*, XLVII, 1962, pp. 192–5.

Nickel-occurrence, products and ores. *Optima*, June 1959, p. 78.

Focus on Malaysia, *New Commonwealth*, April 1971, p. 9.

Simms, G., 'Malayan tin: a rich endowment', *Optima*, June 1968, pp. 74–9.

Sargent, J., 'The Rising Sun in Siberia', *Geographical Magazine*, XLI, pp. 3–7.

Kowalewski, J., 'The Soviet Union's struggle for self-sufficiency in metals', *Optima*, December 1959, pp. 209–15.

Snyder, D. E., 'Ciudad Guayana: Venezuela: Planned Industrial Frontier metropolis', *Geographical Review*, LVII, 2, 1967, pp. 260–62.

Lomas, P. K. and Gleave, M. B., 'Recent changes in the distributions of production in the South African gold mining industry', *Geography*, LIII, 1968, pp. 322–6.

McKinnon, D., 'Minerals, the key to progress in Africa', *Optima*, June 1963, pp. 73–81.

Redrupp, R. J., 'Canada's growing mineral industry', *Optima*, June 1963, pp. 92–100.

Black, R. L., 'Development of South African mining methods', *Optima*, June 1960, pp. 65–77.

MacConachie, H., 'Progress in gold mining over fifty years', *Optima*, September 1967, pp. 130–38.

Rissik, G., 'The growth of South Africa's economy', *Optima*, June 1967, pp. 52–60.

Further Reading

Anderson, T., 'Mineral resources of the Transvaal', *Optima*, March 1967, pp. 34–40.

Gardiner, N., 'The Witwatersrand', *Optima*, March 1968, pp. 20–35.

Dagut, M., 'The South African economy through the Sixties', *Optima*, September 1968, pp. 114–24.

Oppenheimer, H., 'South Africa's growth in the '70's: the problems and the opportunities that lie ahead', *Optima*, December 1971, pp. 154–9.

Pim, J., 'Creating a new landscape for a mining town' (Welkom), *Optima*, March 1956, pp. 26–31.

Welkom – a tailor-made town, *Optima*, March 1961, pp. 12–13.

Le Barrow, W., 'Jagersfontein', *Optima*, June 1971, pp. 85–97.

Wallwork, K. L., 'Some problems of subsidence and land use in the mid-Cheshire industrial area', *Geographical Journal*, CXXVI, 2 June 1960, pp. 191–9.

Hilton, K. J., 'The Lower Swansea Valley Project', *Geography*, XLVIII, 1963, pp. 296–9.

Goodridge, J. C., 'Renewed interest in Cornish tin', *Geography*, XLVII, 1962, pp. 85–7.

Blunden, J. R., 'The Renaissance of the Cornish tin industry', *Geography*, LV, 1970, pp. 331–335.

Goodridge, J. C., 'The tin mining industry: a growth point for Cornwall', *Transactions Institute of British Geographers*, 38, June 1966, pp. 95–103.

255

Glossary

ACIDIC: Description of an igneous rock rich in silica. Made up of predominantly light coloured minerals. Granite is the best known example.

ALLUVIUM: River deposited material – mud, sand, silt. Often makes up great riverside plains.

BASIC: Converse of 'acidic'. Igneous rock poor in silica. Basic rocks are generally darker than acidic rocks. Basalt is an example.

BEDDING PLANE: A surface within sedimentary strata marking an original deposition surface.

BOSS: An intrusive mass forming rounded but rugged topography at the surface.

CAVING: A system of mining in which great blocks are allowed to fall following the removal of their support. The fall breaks the material into a form in which it can be handled and, as working advances, the strata above are allowed to subside.

COUNTRY ROCK: The rock surrounding and penetrated by an igneous intrusion.

DREDGE: A large machine used for collecting and bringing to the surface those minerals worked under water.

FLOTATION: Mineral separation using a froth formed in water by addition of reagents. Finely comminuted minerals float and heavier ones sink.

FLUID MINING: Extraction of minerals as fluids, as with brine, potash, etc.

GANGUE: The waste material associated with the ore in a vein.

GEOPHYSICAL SURVEY: Mineral survey involving tracing and mapping the composition and structure of portions of the earth's surface by a variety of techniques.

GRANITE: Acidic igneous rock formed at considerable depth and therefore coarse grained.

GRAVEL PUMP: Device for breaking up tin-bearing gravels and carrying the material to giant washing trays.

IGNEOUS: Rocks formed by solidification from a molten state, that is from magma.

INTRUSION: Igneous material injected among pre-existing rocks.

JOINT: A surface parting transverse to bedding planes.

Glossary

LATERITE: A product of tropical weathering, involving the deep decomposition of rocks, usually igneous. Contains concentrations of iron and aluminium hydroxides and is an important source of ores such as bauxite, iron ore, nickel and manganese.

LEACHING: In mineral working the process of dissolving minerals from an ore by use of water, acid, cyanide, etc.

LODE: A fissure in country rock filled with metalliferous minerals (*see* vein).

MAGMA: Molten rock material, originating deep within the earth's crust but sometimes injected into the strata above or poured out on to the earth's surface as in volcanoes.

METAMORPHIC AUREOLE: Belt of metamorphosed rocks surrounding an intrusion. Frequently mineral rich.

METAMORPHISM: Change in character of a rock previously either igneous or sedimentary as a result of great pressure, heat or the introduction of new minerals.

PLACER: An alluvial or glacial deposit containing valuable mineral concentrations.

REPLACEMENT DEPOSIT: Mineral deposit formed as a result of removal of original deposits by solution and replacement by other minerals.

ROOM AND PILLAR: Mining of coal or ore by working in rooms separated by narrow pillars of unworked material. The pillars are later extracted successively, the roof being allowed to cave-in behind.

SECONDARY ENRICHMENT: Enrichment of an ore body by addition of material derived from oxidation of overlying mass.

SEDIMENTARY: Rocks formed from consolidation of mechanical sediments or by precipitation from a solution as with sandstone or chalk.

STOPING: Underground mining method involving loosening and removal of ore either by working upwards or downwards.

TECTONIC: Concerning the structure of the earth's crust, the folding and faulting of rock masses. Once neatly summed up as 'the architecture of the earth's crust'.

VEIN: One mineral or several occurring along a line within a country rock. Lode is commonly the term used instead of vein by miners.

WALKING DRAGLINE: Excavator with long boom and usually high capacity bucket capable of removing overburden or mineral and swinging round to drop it well to the rear of the working face.

WEATHERING: Chemical and mechanical action at the surface of the earth leading to decay of rocks.

Index

Mineral Resources

Bell Bay (Tasmania) smelter, Alcan's, 97 *bis*

Belle Plaine, near Regina (Saskatchewan), 38

Benguela Railway, 76, 78–80 *passim*, 82

Bethlehem Steel Corporation (U.S.A.), 58, 60

Bevercotes colliery, East Midlands (U.K.), 40

Biafra war (in Nigeria), 163, 164

Bibliography (for further reading), **250–2**

Bilbao (Spain), 10

Bingham Canyon, Utah (U.S.A.), 116–18; Utah Copper Coy, *117 bis*

Blyth (port in Northumberland), 203–5 *passim*, 207 *bis*

Boers (of South Africa), 217

Boké (Guinea) bauxite ores, 75, 89

Bolivia: 'a poor country', 165; mining in general, 123, 169; tin mining and production, 157–69 *passim*, *165, 166*; nationalization, 166, 167 (*see also* Comibol); difficulties and problems, 164 *bis*, 166, 167–9; silver, 164; mining dynasties of, 165–6; National Revolutionary Movement (M.N.R.), 166, 167, 168; location of tin deposits, 166; labour costs, 167–8; production costs, 167, 168; 'Triangular Operation' (U.S.A., Germany and Western Bolivia), 168; ore exports (concentrates), 168–9; smelting, 169; and present and future Cornish tin mines, 239–40

Bomi Ridge—Monrovia Railway (Liberia), 65

Bomvu Ridge (Swaziland) iron mines, and connecting railway, *64*, 65

Bonneville Power Administration (U.S.A.), 102

Booth, James (aluminium semi-fabricator company), 199, 206

Bootle (Lancashire) smelter, 34

Bougainville (New Guinea) copper mine, 46, 141, *141*

Boulby (N. Yorkshire) potash mine, *232*

Braden field *or* El Teniente (Chile) copper mine, 123–4, *125*, 125, 127 *ter*

Brazil, 15; aluminium, 110 *bis*; iron ore, 75

brick manufacture (U.K.), 31, 32

BRINE (U.K.), 37

Britain (*see also* U.K.): geological surveys, recent, 22; legislation and mining, 202, 227, 231, 232–40;

260

mineral production and imports, 233–4

British Aluminium Company (B.A.C.), 90, 93, 98, 108, 109, 196–205 *passim*, 206

British Aluminium Industry, the New, **196–208**; changing the rules . . . , 196–205; new smelters, 201–2; location, 202–5; smelting, semi-fabrication and regional economic growth, 205–8

British Columbia, 46, 141

British exploitation of Central African minerals, 125

British Insulated Callendar Cables (B.I.C.C.), 139

British South Africa Company, 134, 135

British Steel Corporation, 194

Broken Hill (Zambia) mines, 51, 76

Broken Hill Proprietary (New South Wales): lead, zinc and silver, 70; steel, 216

BROMINE, 23

BRONZE (copper and tin alloy), 112, 113

Brown, Harrison (of California Institute of Technology), 244

Bulawayo—Salisbury (Southern Rhodesia) uplands, 76

Burma: minerals, 53, 236; oil, 53

Butte district of Montana (U.S.A.), 34, 118, 125

Bwana Mkubwa (Zambia) copper mine, 76, 134

C.B.I. (Confederation of British Industry), 194

C.E.G.B. (Central Electricity Generating Board), 201–4 *passim*

C.I.P.E.C. (Intergovernmental Council of Copper Exporting Countries), 142, 143

CALICHE (nitrates), 10; *see also* NITRATES

California gold rush, and subsequent wheat economy, 211

Calumet and Hecla Mining Company, Michigan (U.S.A.), 114

Camborne (Cornwall), 236; Royal School of Mines, 54, 236

Cambridgeshire coprolite, 13

Canada, 223; aluminium industry, 104, 108, 198; asbestos, 30; gold, 33, 222; nickel, 30

Canadian: Eastern Area, 67; Geological Survey, Report of (1894), *quoted*, 65; mineral development and transport, 65–7; National Railway, 67; Pacific

Mineral Resources

North American (1966), *118–19*; mine, smelter and refinery production, and consumption, of leading countries (1968) (*graph*), *121*; copper-nickel deposits, Norilsk (U.S.S.R.), 39; ore, and smelted and refined copper, production of, and consumption of refined copper (1967), *137*; prices, 48; prices on London Metal Exchange and of U.S. producers (1960–70 inclusive), *138*; producers in the capitalist world, leading (1929–69), *132*; production and consumption, world (1955, 1960, 1965 and 1969), *113*; results, *quoted*, 115–16; richness of ores, 114; smelting, 51; sources, *see below*; substitutes, 112, 138–9; uses of, 113–14; output, 154

COPPER (*sources*): Africa, Central, 34, 115, 129–36; Andes, the, 115; Anglesey, 58, 112; Atacama desert, 124, 128; Australia, 17; Chile, 114, 120 *bis*, 123–9, 132 *bis*; Congo, 120, 129–32; Cornwall, 114; Dzhezkazgan (U.S.S.R.), 115; Kazakhstan (U.S.S.R.), 115 *bis*; Latin America, 34, 114, 120–9; Mexico, 120; North America, 34, *118–19*, 223; Northern Rhodesia, 76, 120, 132; Pacific Ocean, 27; Peru, 120 *ter*; Spain, 114; Third World, **120–36**; Transvaal, 214–15; U.S.A., 34, 114, 115, 116–20, *119*; U.S.S.R., 114, 115, 120, 181; Western Europe, 34; Zambia, 39, 51, 76, 120 *bis*, 129, 132–6

Copper Belt (Northern Rhodesia, Zambia and Congo), 39, 51, 76 *ter*, 77, *78*, 79, 80, 82, 130, 132–6 *passim*; watering problems in, 133 *ter*

COPROLITE (fossilized excrement), 13, 230

Corby area (Northamptonshire), 32 *bis*, 33

Cornish tin: mining (1960–71), *238*; smelting, 85

Cornwall, 49; mineral ports, 57, *58*; tin mining revival, 154, 157, 171; old mining areas, 209–10, 236–7, *237*; unemployment, 234, 237; copper and tin, 236–40; new projects, 236–40, *238*; difficulties of revival, 237–9; old mines re-working, 239; a new and an old mine, 239; developments in the near future, 239; and their effects on Bolivia, 239–40

262

Corporacion Minera de Bolivia, *see* Comibol

costs and prices of minerals, 47–9

CRYOLITE, 86 *bis*

Cuajone (Peru) copper, 45, 46, 114–15, 141, 143

Cuba nickel, 147

Daldyn (U.S.S.R.) diamond region, 21

Dampier (port in Western Australia), 74 *quin*; iron ore, 74, *75*, 215; shipping, 74, 215

Dampier – Mount Tom Price railway, 74, 215

Dartmoor (Devon), 239

Darwin (Australia, N.T.) uranium, 20

De Beers Mining Company (South Africa), 217

Derbyshire lead and zinc mines, 154

Development Areas (U.K.), 49, 199, 202, 205

Development Decade, Second, 243

Development of Capitalism in Russia, The (Lenin, 1899), *quoted*, 175–6

DIAMONDS, 9, 21, 31, 217

Dolcoath mine (1904), Redruth (Cornwall), 236, *237*

Donbass (U.S.S.R.), 176; coal, 176

Dredge mining, Malaysia, *161*

dredges, 10; Malayan tin-mining, *161*, 162, 163

Drilling iron ore at Kiruna (Sweden), *38*

Ducktown, Tennessee (U.S.A.), smelter, 34

dumper trucks, high-capacity, 42, *42*, 135

Duncan, Sir Val, 55

Durham, West, 234

Dzhezkazgan (U.S.S.R.): copper, 115, 192; smelter, 192–3

E.E.C. (European Economic Community *or* Common Market), 198, 199

East Midlands (of England): coalfield, 19, 40; conurbation, 230; ironstone, 227, 230

Eastern Siberia, *see* Siberia, Eastern

economic development and minerals, *see* Mining and Economic Growth

economies and economics, mining, *see* Mining Economics

Economist, The: on the New British aluminium industry, *quoted*, 201

Eire: legislation and mining, 233; mining, 154; zinc and other base metals, 154

Mineral Resources

264

Mineral Resources

Mineral Resources